W9-AZF-934

PERGAMON INTERNATIONAL LIBRARY
of Science, Technology, Engineering and Social Studies
The 1000-volume original paperback library in aid of education, industrial training and the enjoyment of leisure
Publisher: Robert Maxwell, M.C.

ENERGY AROUND THE WORLD

AN INTRODUCTION TO ENERGY STUDIES

Global Resources, Needs, Utilization

THE PERGAMON TEXTBOOK INSPECTION COPY SERVICE

An inspection copy of any book published in the Pergamon International Library will gladly be sent to academic staff without obligation for their consideration for course adoption or recommendation. Copies may be retained for a period of 60 days from receipt and returned if not suitable. When a particular title is adopted or recommended for adoption for class use and the recommendation results in a sale of 12 or more copies, the inspection copy may be retained with our compliments. The Publishers will be pleased to receive suggestions for revised editions and new titles to be published in this important International Library.

Other Pergamon Titles of Interest

ARIZONA STATE UNIVERSITY LIBRARY	Solar Energy Index
DIXON & LESLIE	Solar Energy Conversion
EGGERS LURA	Solar Energy for Domestic Heating & Cooling Solar Energy in Developing Countries
HALL	Solar World Forum (4 volumes) Biomass for Energy in the Developing Countries
HELCKE	The Energy Saving Guide
HOWELL	Your Solar Energy Home
JAGER	Solar Energy Applications in Dwellings
MARTINEZ	Solar Cooling & Dehumidifying
McVEIGH	Sun Power, 2nd Edition
O'CALLAGHAN	Building for Energy Conservation Energy for Industry Design & Management for Energy Conservation
OHTA	Solar Hydrogen Energy Systems
PALZ & STEEMERS	Solar Houses in Europe: How They Have Worked
REAY	Industrial Energy Conservation, 2nd Edition
REAY & MACMICHAEL	Heat Pumps
SECRETARIAT FOR FUTURES STUDIES	Solar versus Nuclear: Choosing Energy Futures
TAHER	Energy: A Global Outlook

Pergamon Related Journals *(free specimen copy gladly sent on request)*

Energy
Energy Conversion & Management
International Journal of Hydrogen Energy
Journal of Heat Recovery Systems
OPEC Review
Progress in Energy & Combustion Science
Solar Energy
Sun at Work in Britain
Sun World

ENERGY AROUND THE WORLD

AN INTRODUCTION TO ENERGY STUDIES

Global Resources, Needs, Utilization

By

J. C. McVEIGH

M.A., M.Sc., Ph.D., C.Eng., F.I.Mech.E., F.Inst.E., M.I.Prod.E., M.C.I.B.S.

Head of Energy Studies, Brighton Polytechnic

PERGAMON PRESS

OXFORD · NEW YORK · TORONTO · SYDNEY · PARIS · FRANKFURT

U.K.	Pergamon Press Ltd., Headington Hill Hall, Oxford OX3 0BW, England
U.S.A.	Pergamon Press Inc., Maxwell House, Fairview Park, Elmsford, New York 10523, U.S.A.
CANADA	Pergamon Press Canada Ltd., Suite 104, 150 Consumers Road, Willowdale, Ontario M2J 1P9, Canada
AUSTRALIA	Pergamon Press (Aust.) Pty. Ltd., P.O. Box 544, Potts Point, N.S.W. 2011, Australia
FRANCE	Pergamon Press SARL, 24 rue des Ecoles, 75240 Paris, Cedex 05, France
FEDERAL REPUBLIC OF GERMANY	Pergamon Press GmbH, Hammerweg 6, D-6242 Kronberg-Taunus, Federal Republic of Germany

Sci Engr.
TJ
163
.2
.M388
1984

Copyright © 1984 J. C. McVeigh

All Rights Reserved. No part of this publication may be reproduced, stored in a retrieval system or transmitted in any form or by any means: electronic, electrostatic, magnetic tape, mechanical, photocopying, recording or otherwise, without permission in writing from the publishers.

First edition 1984

Library of Congress Cataloging in Publication Data
McVeigh, J. C.
Energy around the world.
(Pergamon international library)
Includes index.
1. Power resources. I. Title.
TJ163.2.M388 1984 333.79 84–14878

British Library Cataloguing in Publication Data
McVeigh, J.C.
Energy around the world.—(Pergamon international library)
1. Power resources
I. Title
333.79 TJ163.2
ISBN 0–08–031649–2 Hardcover
ISBN 0–08–031650–6 Flexicover

Printed in Great Britain by A. Wheaton & Co. Ltd., Exeter

A find of energy in Nature means an addition to the general wealth, a postponement of the day of bankruptcy, which each new invention of science, on the other hand, brings nearer.

Professor Sir Frederick Soddy
Matter and Energy 1912

PREFACE

The aim of this book is to provide a balanced overview of the main issues in this complex interdisciplinary subject, so that the readers will be able to make their own informed contributions to the energy debate. The emphasis is on the availability of energy resources and how the future pattern of world energy supply and demand could develop. It is intended to provide a general introduction to energy studies for sixth formers and first year degree students in a wide range of disciplines including science, engineering, architecture, building, economics and geography. It should be particularly useful for teachers and others who need access to reliable basic energy data and statistics, suggestions for more detailed study in particular topics and ideas for suitable exercises and projects.

When I wrote the preface to my first book Sun Power, an introduction to the applications of solar energy, in 1976 I pointed out that we were living at a time when there was a greater awareness of the energy problems facing the world than at any other period in history

...the growth in energy consumption which has been experienced for many years cannot continue indefinitely as there is a limit to our reserves of fossil fuel. Solar energy is by far the most attractive alternative energy source for the future...

Since then I have been invited to teach and lecture in many schools, colleges, Polytechnics and Universities. At first this work was limited to solar energy applications, but it soon widened to include all the alternative energy sources and to place these alternatives in the context of present and possible future trends in the production and consumption of the conventional fossil fuels and nuclear power. More recently an increasing number of sixth form and first year degree courses in the UK have included energy topics as part of the curriculum in subjects as diverse as the school 'A' level course in 'Social Biology' and the engineering degree subject 'The Engineer in Society'. The book grew from my lecture notes and other material specially prepared for these energy studies lectures and seminars.

While a basic scientific knowledge would be an advantage for a complete understanding of each topic, I have followed the pattern adopted in Sun Power of keeping the theoretical treatment to a minimum. This enables a broad appreciation of the subject to be obtained in a fairly short period. Tables of statistics have also been kept to a minimum by using carefully selected abstracts backed up with original analyses and illustrations showing the main trends. The first chapter gives a brief introduction to the nature of energy with a review of some basic definitions and principles. This is followed by a historical chapter which traces the development of the main energy resources and then discusses the period of very rapid growth in energy production and consumption from the 1950s. This chapter also examines trends in world population and the growth in the economies of several selected countries. The following seven chapters are arranged to reflect their relative importance in supplying the world's energy in the 1980s - oil-coal-gas-hydropower and biomass-nuclear-solar-other renewables. The final chapter looks at forecasting and some other possible technical solutions and resources. It also gives an overview of the main issues and attempts to outline some future possibilities. Each chapter contains a brief summary of the most important points, a list of references, suggestions for further reading and a set of original exercises. Some of the exercises require a basic scientific knowledge and include simple mathematical models. Answers to most of these numerically based questions are given. Other exercises have been chosen to bring out broader social, economic, political and environmental issues. In some cases a complete answer will require additional reading and statistical research.

I have recently had the opportunity of visiting and teaching in a number of overseas institutions, including the Asian Institute of Technology, Bangkok, the Building Research Institute in Bucharest, Romania, the University of Khartoum, Sudan, the University of Moratuwu, Sri Lanka, the National Institute for Higher Education, Limerick, Republic of Ireland, the University of Sana'a in the Yemen Arab Republic and Zhejiang University in the People's Republic of China. I would like to thank all my friends and colleagues in these institutions for their kindness and hospitality and hope that they will see something of their philosophy reflected in this book. The importance of fuel wood in the lives of millions of people in the developing countries is easily overlooked in western industrial society and I am particularly grateful to Dr Lesley Brattle, formerly of the University of Surrey, for allowing me to include her original material in Chapter 6, based on her field studies on the problems of cooking and the use of fuel wood. I would also like to thank John Millar, the founder of the World Solar Power Foundation, for his encouragement and for many discussions on the wider implications of energy and the environment. My teaching experience at school level has been greatly enhanced through the kindness of John Hunt, Headmaster of Roedean School, Brighton, who invited me to lecture and teach over the whole range of topics covered in the book to successive groups of sixth formers. The development of several ideas in the text owes much to their interest and enthusiasm. I would also like to thank Professor Peter Dunn of Reading University, Professor Douglas Probert of Cranfield Institute of Technology and Simos Yannas of the Architectural Association for inviting me to lecture and give seminars to their post-graduate energy groups. This experience is also reflected in the text.

I wish to thank all those who have helped me to gain access to a wide range of publications and other material. These include the Press and Publicity Offices of the National Coal Board, British Gas, the Central Electricity Generating Board, the South of Scotland Electricity Board, the Department of Energy, the Shell Transport and Trading Company and the Esso Petroleum Company. I wish to thank Her Majesty's Stationery Office for permission to use abstracts from statistical data prepared by the Department of Energy. I would like to thank Dr Ian Blair, of the Nuclear Environmental Branch at Harwell who studied the final draft of this chapter and made some valuable suggestions. I acknowledge with thanks permission from Nuclear Power International to include extracts from their nuclear power station statistical performance data. I owe a particular debt to Gerald Foley who, through his book 'The Energy Question', introduced me to the works of Professor Sir Frederick Soddy. In 1921 Soddy warned the world about the folly of using up the 'capital' energy stored in the fossil fuels instead of looking to the continuous 'revenue' of energy received from the sun. It is sad that it took nearly half a century for his ideas to be rediscovered. Bruce

Andrews of the Financial Times Business Newsletters has also provided encouragement over many years and drawn my attention to a number of useful information sources. The statistical data provided by Tony Scanlon and his predecessor Ken Inglis of the British Petroleum Company has been invaluable and I am also very grateful to Ken Inglis for his helpful review of the final draft of the Petroleum Chapter. I would like to thank Bernard McNelis for providing Figure 8.7 and the National Coal Board for Figure 4.1 which shows their chairman, Mr. Ian MacGregor, studying working conditions at a coal face. The entire text was prepared in the 'camera-ready' form by Vibes Limited of Brighton and I would especially like to thank Angela Barden for her editorial assistance.

I would like to thank Ewbank Preece of Brighton for their encouragement and support for my renewable energy work and I acknowledge the continued support from the Director and Council of the Brighton Polytechnic. I owe a special word of thanks to my colleagues in the Learning Resources Centre at the Brighton Polytechnic who traced and obtained a large number of references from other libraries.

It was again a great pleasure to work with Jim Gilgunn-Jones and his team at Pergamon Press and I take this opportunity of expressing my thanks to him and to all who have been involved in the production of the book. Finally I would like to acknowledge the encouragement, patience and support which I received from my wife and family. Without their help the book could not have been written.

Cleland McVeigh

Brighton
June, 1984

CONTENTS

1. WHAT IS ENERGY?

Introduction

What is energy? Modern ideas about the nature of energy were probably first developed in the seventeenth century by the English scientist Sir Isaac Newton. His Laws of Motion led to the basic concept that energy is neither created or destroyed. The principles of energy conversion were gradually developed from the study of the early steam engines at the start of the Industrial Revolution. Today, terms such as 'energy consumption' or 'energy shortage' are commonly used. Although these terms are not correct, they are used in this text with a suitable definition. Energy is not 'consumed'. It is changed from a form in which it is not being useful to another form of energy which can be used. A simple example is the transformation of the stored chemical energy in coal to electricity. All the stored energy can be accounted for. Only a minor proportion, about 30%, appears as electricity where it is used. Some appears at the power station as hot exhaust gases or as hot water. The rest appears as heat transferred to the building, its surroundings or in the transmission wires.

In this chapter some of the basic definitions and principles are reviewed. These may be familiar to those with a scientific background and some terms will be defined rather briefly, with reference to more rigorous texts.

Energy

The chapter commenced with the question "What is energy?" Energy can be classified in various ways. It can be considered as an abstract idea. It is a concept which is useful in describing physical changes in the real world. All the various forms of energy can be linked in one common definition:

Energy is the capacity to do work

This can be written in other ways. For example "Energy is the ability to apply a force through a distance".

Some other common forms of energy which will be discussed include thermal energy or heat, kinetic energy, potential energy and chemical energy. Three main classifications are often used in energy statistics. These are Primary, Delivered and Useful or End Use Energy.

Primary Energy is often considered to be the energy available in any fossil fuel such as oil or coal before it has been processed or transmitted. Primary energy is also energy potentially available from the natural environment. Solar energy, wind or wave power are examples.

Delivered Energy is the energy actually supplied to the consumer. It is always less than the original primary energy because of transmission and conversion losses.

Useful or End Use Energy is the energy available to the consumer. This can take many forms including heat and light or mechanical work.

Work

The mechanical work done, W, when a force F, moves a body a distance x in the direction of the force as given by the equation:

$$W = Fx \tag{1.1}$$

In the SI system, described later in this chapter, the unit of length or distance is the metre (m), the unit of force is the newton (N) and work is measured in joules (J). One joule is exactly one newton-metre. All physical quantities can be as a function of three basic independent variables, usually length, mass and time. Mass is a quantity of matter. Force is sometimes used instead of mass in some definitions.

From Newton's Second Law of Motion:

$$\text{Force} = \text{rate of change of momentum}$$

$$F = \frac{d}{dt}(\text{mass x velocity})$$

$$F = \frac{d}{dt}\left(m\frac{dx}{dt}\right)$$

$$F = m\frac{d^2x}{dt^2} \tag{1.2}$$

or Force is the product of mass and acceleration

Power

Power is defined as the rate of work or the rate at which any energy conversion takes place. The fundamental unit of power is the watt (W). One watt is one joule per second.

Kinetic Energy

This is the energy due to the motion of a body of a given mass. The kinetic energy of a body of mass m and velocity u is $\frac{1}{2}mu^2$.

Potential Energy

A more precise term is 'gravitational potential energy', as it is the energy possessed by a body at some height above a fixed datum line. The body has the potential for the gravitational force to do work on it. The potential energy of a body of mass m at a height z above the datum line is mgz, where g is the gravitational acceleration normally taken as 9.81 ms^{-2}.

Chemical Energy

Chemical Energy is the energy stored in a substance due to its chemical compostion. Chemical energy is usually released by combustion.

Internal Energy

The intrinsic energy of a body or a fluid at rest is known as its Internal Energy. It can always be measured by referring to other measurable properties such as temperature and pressure.

Heat

Heat is a form of energy. It is transferred from one body to another body at a lower temperature because of the temperature difference between the two bodies. Both heat and work can be thought of as energy in transition. They are not contained or possessed by the body. There are, in general, three ways in which heat may be transfered - by conduction, convection or radiation. Most practical applications involve a combination of all three modes of heat transfer.

Conduction

Conduction is the transfer of heat from one body, or part of body, to another part of the body or to another body in physical contact with it.

Convection

Convection is the transfer of heat within a fluid through the internal movement of the fluid. Convection can be natural and caused by differences in density resulting from differences in temperature, or forced and caused by a machine such as a pump or fan.

Radiation

Electromagnetic radiation is emitted by all bodies. This radiant energy is proportional to T^4, where T is the absolute temperature (K). It also depends on the colour and surface texture of the body. The transfer of heat by radiation obeys the Stefan-Boltzman radiation law:

$$Q = \sigma T^4 \tag{1.3}$$

where σ, the Stefan-Boltzman constant, = $5.669 \times 10^{-8}\,Wm^{-2}$, and Q is the total energy emitted at all wavelengths by an 'ideal' body. The ideal body is called a 'black body' and is a perfect emitter and absorber. The radiation emitted by a real body is always less than the ideal. The ratio of the actual radiation emitted to the ideal is known as the emittance.

A second important relationship concerns the intensity of energy distribution. As the temperature of emission falls, the wavelength corresponding to the maximum emission of radiant energy, λ_m, increases according to the following relationship:

$$\lambda_m = \frac{C}{T} \tag{1.4}$$

where C is approximately 3000 when λ_m is measured in microns. This is known as Wien's displacement law. For example, the maximum intensity of radiation from a body at 60°C occurs at about 9 microns, compared with about 0.5 microns for incoming solar radiation which consists of wavelengths ranging from the visible (0.38 - 0.78 μm) into the near infrared (0.78 to just over 2.00 μm).

This is the explanation for the well-known 'greenhouse effect'. The short wavelengths of the incoming solar radiation can pass through glass, with up to 90% being transmitted directly. In the wavelength range above 3 microns the transmittance of glass is practically zero. The longwave radiation from bodies at temperatures less than 700°C cannot pass directly back out through the glass.

Nuclear energy, electricity, magnetism, sound and light are all examples of different forms of energy. Mass is also a form of energy. The relationship between energy and mass was first developed by Albert Einstein with the equation

$$E = mc^2 \tag{1.5}$$

where E is the energy released (J), m is the mass converted into energy (kg) and c is the velocity of light (3×10^8 ms^{-1}). This equation and others dealing with the behaviour of atoms and molecules will not be discussed in this text.

The First Law of Thermodynamics

The principle of the conservation of energy was well established by the early nineteenth century. When it was applied to heat energy and mechanical work, it was known as the first law of thermodynamics. It is often expressed as follows:-

In any enclosed system, the change of internal energy of the system is equal to the net amount of heat transferred to the system (Q) less the net external work done by the system (W). If E_1 and E_2 represent the initial and final internal energy of the system then:

$$Q - W = E_2 - E_1 \tag{1.6}$$

To obtain a continuous work output it is necessary to bring the system back to its original state, i.e. it must pass through a cycle of operations.

Using the first law, the efficiency of any energy transformation or conversion system can be simply defined as

$$\frac{W}{E} \times 100\% \tag{1.7}$$

where W is the desired energy or work provided by the system and E is the net energy input. This is known as the first law efficiency. It is a useful method for comparing the performance of energy conversion systems of any particular type. For example, the performance of various gas-fired water heating boilers could be compared by measuring the amount of gas needed by each boiler to raise a given amount of water through the same temperature rise. In every case some of the original stored chemical energy in the gas will be dissipated to the surroundings. The dissipated energy, in the form of exhaust gases or low temperature heat, is no longer easily available for further energy transformation.

However, the first law has very important limitations. It does not describe the changes in the quality of the energy or its capacity to do work. These are described in the second law of thermodynamics. The second law pcints out, for example, that two litres of water at 35°C could not transform itself into one litre of hot water at 60°C and one litre at 10°C unless some other energy transfers occured. The reverse process of mixing one litre at 60°C with one litre at 10°C to produce two litres at 35°C is possible (provided no energy transfer takes place with the surroundings). Although energy has been conserved there has been a degradation in the usefulness or availability of the energy. Try filling your bath with a large amount of 35°C water rather than a small amount of 60°C water!

The Second Law of Thermodynamics

The second law can be stated in a number of different forms. One form is as follows:-

It is impossible to construct a device which will operate in a cycle and perform work while exhanging energy in the form of heat with a single reservoir. To obtain a work output, two reservoirs are always needed. The higher temperature reservoir is often called a source and the lower temperature reservoir a sink. In another form the second law states that heat transfer can only take place from a hotter to a cooler body.

In equation 1.6 the net amount of heat transferred, Q, is composed of two parts. Q_1 is the heat supplied at a higher temperature than Q_2, which is the heat rejected at a lower temperature. The system then becomes the simple theoretical heat engine shown in Figure 1.1.

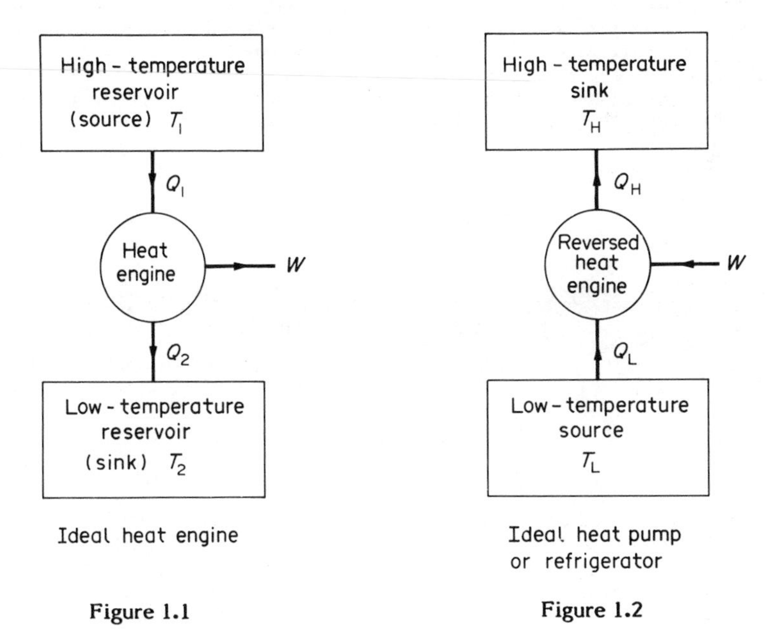

Figure 1.1 **Figure 1.2**

The efficiency of the cycle is the net work output W divided by the heat input Q_1:-

$$\eta = \frac{W}{Q_1} \tag{1.8}$$

As the operation is cyclic, $W = Q_1 - Q_2$, and the efficiency can also be expressed as:-

$$\eta = \frac{Q_1 - Q_2}{Q_1} \text{ or } 1 - \frac{Q_2}{Q_1} \tag{1.9}$$

If the absolute temperature of the source is T_1, and the sink T_2, then the cycle efficiency becomes:-

$$\eta = 1 - \frac{T_2}{T_1} \tag{1.10}$$

This is known as the ideal cycle or Carnot efficiency, after the Frenchman Sadi Carnot who was the first to develop these concepts in 1824. This theoretical Carnot efficiency will always be greater than the efficiency of any real engine working between the same temperature limits. There are various reasons for this, the main ones being energy losses due to friction between the moving parts and the need for a temperature difference between the source and the engine, and between the engine and the sink, so that heat transfer can take place. In practice it is very valuable to use the Carnot efficiency on a comparative basis, bearing in mind that at best the efficiency of a real engine will be about two-thirds of the Carnot efficiency.

If the processes in Figure 1.1 are completely reversed, then the reversed heat engine has work done on it, W, and transfers heat Q_L from the low temperature source at T_L. The heat transferred to the high temperature sink, Q_H is then given by

$$Q_H = W + Q_L \tag{1.11}$$

In this form, the system acts as an ideal heat pump. The amount of heat delivered, Q_H can be considerably greater than the work W. The ratio Q_H/W is known as the coefficient of performance (CoP) or performance ratio of the heat pump. The system can also be considered to be a refrigerator, but in this case the CoP is Q_L/W, as only the heat removed at the lower temperature is of importance. This is shown in Figure 1.2

Entropy

One of the consequences of the second law is that as energy flows from the higher temperature reservoir it is also dissipating to the universe. The low temperature reservoir is also increasing its energy from the universe. This led the German physicist Clausius, some forty years after Carnot, to define a quantity called entropy. In any process the change in entropy is equal to the change in heat divided by the absolute temperature. The gain in entropy of a system is a measure of the unavailability of heat for transformation into mechanical work. Entropy is a difficult concept to grasp. The first and second laws of thermodynamics could be summarised by stating that although the total energy content of the universe is constant, its total entropy is always increasing. In other words, as time passes less of the energy in the universe is available for transformation into work. Eventually everything will reach the same temperature. All the energy in the universe will be in the form of low temperature heat. Lord Kelvin called this the "heat death of the universe".

Available Work

Available work is a concept which describes the quality of any energy transformation as well as its quantity. It measures the potential of a thermodynamic system to do work (2). An efficiency which shows performance relative to what is theoretically possible subject to the constraints of the second law of thermodynamics is the second law efficiency, ε. This is defined as

$$\frac{W_m}{W_c} \times 100\% \qquad (1.11)$$

where W_m is the least available work required for a task and W_c is the available work actually used in carrying out the task. The first step in calculating ε is to define the task. If the task involves work W, then $W_m = W$. With heat, the least available work required is surprisingly less than the amount of heat delivered. This is because the ideal process would involve the use of heat pump. For example, if it is required at a temperature T_H with the surroundings at temperature T_L then

$$W_m = Q_H(1 - \frac{T_L}{T_H}) \qquad (1.12)$$

Available Energy

Available energy is the maximum useful work that can be performed by a system interacting with the environment[1]. The maximum available energy from a reservoir at temperature T_1, providing heat Q_1 in surroundings at temperature T_2, is given by

$$X = Q_1 (1 - \frac{T_2}{T_1}) \qquad (1.13)$$

where X is also known as the 'exergy'. Exergy is that part of the energy which is available for doing useful work.

A further consideration of available energy, or exergy, suggests that the energy source should always be matched as closely as possible to the end use. If this is done, the output exergy from any system can be substantial and available for other suitable lower grade applications. A simple example is the concept of combined heat and power (CHP). The high grade chemical energy from the fossil fuel is used to generate electrical power. Most of the output exergy from this process is in the form of hot water, some 30 to 50°C above the temperature of the surroundings. This hot water can be used to provide space heating for nearby buildings. The first law efficiency of the process is often doubled.

Units and Conversion Factors

The International System of Units, or Système International d'Unités, (SI), has been adopted in many countries since 1960, including the United Kingdom. The six basic physical unit quantities are given in Table 1.1

Table 1.1

Basic quantity	Unit	Symbol
Length	Metre	m
Mass	kilogramme	k
Time	second	s
Temperature	kelvin	K
Electric Current	ampere	A
Luminous Density	candela	cd

All other physical quantities are derived from these. The following additional definitions and conversion factors may also be used:

Length

1 millimetre (mm)	0.0393701 inch (in)
1 metre (m)	3.28084 feet (ft)

Area

1 square centimetre (cm^2)	0.155000 in^2
1 square metre (m^2)	10.7639 ft^2
1 hectare = $10^4 m^2$	2.4710 acres

Volume

1 cubic centimetre (cm^3)	0.0610237 in^2
1 cubic metre (m^3)	35.31477 ft^3
1 litre (l) (or one thousand cubic centimetres)	1.75985 UK pints
1 imperial gallon (UK)	4.54596 litres
1 US gallon	3.78531 litres

1 barrel = 42 US gallons = 34.97 UK gallons = 159.00 litres
1 barrel of oil = 0.136 tonnes and has an equivalent heat content of 5.694×10^9 J

Weight

1 kilogram (kg)	2.20462 lb
1 tonne (10^3kg)	0.9984207 ton (UK)

1 ton (UK) or Statute or long ton = 1.120 short tons

Force

1 newton (N)	0.2248 lbf

Pressure

1 pascal (Pa)	1 Nm^{-2}
1 bar = 10^5 Pa	14.50 lbf in^{-2}
1 lbf in^{-2} (one pound per square inch or psi)	6.89476 kPa
Atmospheric pressure = 14.70 lbf in^{-2}	101.325 kPa

Heat, Energy and Power

The British Thermal Unit (Btu) is the amount of heat required to raise 1 lb of water through 1 degree Fahrenheit.

The Calorie is the amount of heat required to raise 1gm of water through 1 degree Celsius (Centigrade).

The Joule (J) is commonly used as a mechanical unit of heat.

1 calorie (cal) = 4.1868J

The fundamental unit of power is the Watt (W).

The Langley is a unit of energy frequently used in solar radiation work and is equivalent to 1 calorie/cm^2.

1 watt = 1 Joule per second = 0.001341 horsepower (hp)
1 horsepower = 745.7 Watts = 550 ft lb per second
1 kilowatt (kW) = 1.34 horsepower
1 Btu = 1.05506×10^3J = 778.169 ft lb
1 kilowatt hour (kWh) = 3.6×10^6J or 3.6 MJ = 3.41213×10^3 Btu
1 therm = 10^5 Btu = 29.3 kWh = 1.05506×10^8J
1 k cal/m^2 = 0.3687 Btu/ft^2 = 1.163 Wh/m^2
1 W/m^2 = 3.6 kJ/m^2/h = 0.317 Btu/ft^2/h
1 Btu/ft^3 = 3.726×10^4 J/m^3

Abbreviations in SI Units

Kilo	k	10^3	milli	m	10^{-3}
Mega	M	10^6	micro	μ	10^{-6}
Giga	G	10^9	nano	n	10^{-9}
Tera	T	10^{12}	pico	p	10^{-12}

For example, 1 megawatt = 1MW or one thousand kilowatts or one million watts.

Temperature Conversions

1 scale degree Celsius (Centigrade), °C = 1.8 scale degrees Fahrenheit, °F.

°C = 5/9 (°F - 32) or °F = 9/5 °C + 32

Temperature on the Kelvin Scale is obtained by adding 273.15 to temperatures on the Celsius Scale. Note that 40°C-20°C = 20K.

Conversion Factors for the Fossil Fuels

Conversion factors from a tonne of coal or oil to other units, such as therms or GWh electricity produced, will always be approximate. The mean calorific value of each fuel varies according to the proportions of the different varieties. For example, in the 1979 Digest of UK Energy Statistics, the calorific values of 26 different types of coal range from 224 to 316 therms per tonne, and of 12 different types of petroleum from 400 to 496 therms per tonne. Statistical reviews always give the conversion factors they use, but these can vary from year to year. Note that "electricity generated" is the approximate amount of electricity that could be produced using that amount of fuel, while "electrical energy" is the actual energy equivalent of the fuel. The "electricity generated" figure will be about 30% of the "electrical energy" figure. For all UK Energy Statistics, the approximate conversion figures given by the Department of Energy in 1983 were as follows:-

Coal — 1 million tonnes is equivalent to 0.6 million tonnes of oil or 250 million therms (2.5×10^{13}Btu) or 7.35 TWh electrical energy or 2 TWh elecricity generated, or 2.637×10^{16}J.
(m.t.c.e. is "million tonnes of coal equivalent")

Oil — 1 million tonnes is equivalent to 1.7 million tonnes of coal or 425 million therms (4.25×10^{13}Btu) or 12.5 TWh electrical energy or 3.6 TWh electricity generated, or 4.484×10^{16}J. The definition given in the BP Statistical Review of World Energy 1982 equates 1 million tonnes of oil with 3.97×10^{13}Btu and 1.5 million tonnes of coal equivalent, or 4.0 TWh electricity generated, or 4.115×10^{16}J.
(m.t.o.e. is "million tonnes of oil equivalent")

Natural Gas — 1 million cubic feet is equivalent to 40 tonnes of coal or 10 thousand therms. A common unit is a trillion cubic feet (Tcf) which is 10^{12} cubic feet, and equivalent to 40 million tonnes of coal (1.05×10^{18}J). 1 cubic foot is the equivalent of 1000 Btu or 1.05×10^{6}J. This gives 1 cubic metre as the equivalent of 3.71×10^{7}J. The BP Statistical Review prefers 1 cubic metre as the equivalent of 9000 k cal, which is 3.77×10^{7}J.

Growth

When a quantity, y, increases by a constant amount with corresponding equal increases in a second variable t, the growth is linear. This is shown by the simple equation:

$$y = y_o + ct \qquad (1.14)$$

Where y_o is the initial value of the quantity when t=0 and c is a constant.

While the early growth rates in the consumption of the main fossil fuels have appeared to be linear for quite long periods, a detailed examination shows that this is not the normal pattern. The rate of increase often starts to increase with time.

Compound Growth

If the quantity y increases annually by an equal rate r, where 1>r>0, then at the end of the first year $y_1 = y_o(1+r)$
where y_o is the initial value of the quantity and y_1 the value at the end of the year. The general equation after n years is

$$y_n = y_o(1+r)^n \qquad (1.15)$$

Example 1.1 — If the annual growth rate is 16%, find the values of y after 1, 5, 10, 20, and 50 years, if $y_o = 1$

Solution: The values are given by

$$y = [1 + 0.16]^n$$

The complete answer is as follows:

n	1	5	10	20	50
y	1.16	2.10	4.41	19.46	1670

One common use of compound growth is to find how long it takes a given quantity y_o to double at a particular annual growth rate r.

This is found by solving the equation

$$2 = (1 + r)^n$$

The doubling times for several values of r are given in Table 1.2.

Table 1.2
Doubling times for annual growth rates, r

r	0.01	0.02	0.04	0.06	0.08	0.10
doubling time in years	69.66	35.00	17.67	11.90	9.006	7.273

Exponential Growth

If y is a function of t such that the rate of growth, $\frac{dy}{dt}$ is proportional to y, then as y increases, the rate of growth also increases. In this case:

$$\frac{dy}{dt} = ct \tag{1.16}$$

where c is a constant, and the solution is

$$y = y_o e^{ct} \tag{1.17}$$

The exponential function, e, is the unique function whose derivative is equal to the function itself

$$\frac{d(e^t)}{dt} = e^t \tag{1.18}$$

Its value is 1 when t = 0.

The relation between the compound growth expression $y = y_o(1 + r)^n$ and the exponential function $y = y_o e^{ct}$, where c is a constant, is found by putting n = t = 1 .

Then $(1 + r) = e^c$ or $\ln(1 + r) = c$

Example 1.2 Using the data in Table 1.3 show that the growth in output rate is approximately exponential and find a suitable exponential and compound growth relation.

Table 1.3
Annual output of the U.K. coal industry (millions of tons)

Date	Annual Output (Q)
1857	65.4
1862	81.6
1867	104.5
1872	123.5
1877	134.2
1882	156.5
1887	162.1
1892	181.8
1897	202.1
1902	227.1
1907	267.8

Solution: For the first thirty years, the increase in output seems to be approximately linear, but in the final period the annual rate of increase is seen to rise as shown in Figure 1.3.

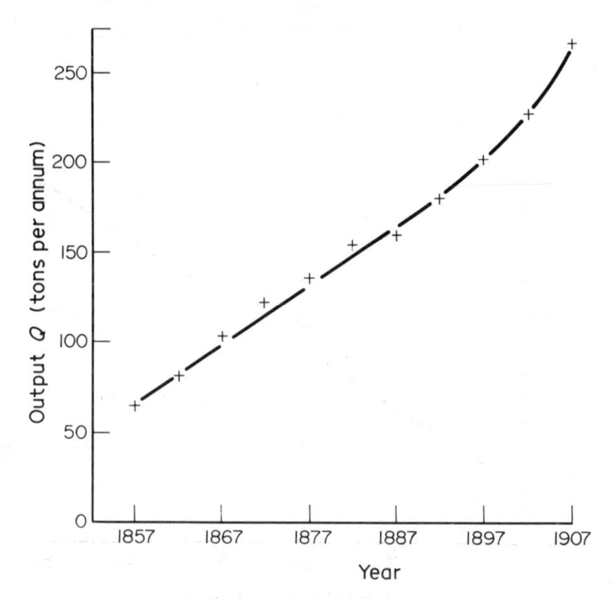

Figure 1.3

Assuming that the relationship is exponential and of the form $Q = Q_o e^{ct}$, the constant Q_o and c can be found by plotting ln Q against t. This is shown in Figure 1.4 where the equation of the best straight line is $\ln Q = 4.36 + 0.0248t$. The equation then becomes

$$Q = 78.26e^{0.0248t}$$

The full results of this approximation are given in Table 1.4, where the values of Q' are calculated from the equation and compared with real values.

Table 1.4

Date	Annual output Q	lnQ	Q'	Q-Q'
1857	65.4	4.18	78.3	-12.9
1862	81.6	4.40	88.6	-7.0
1867	104.5	4.65	100.3	+4.2
1872	123.5	4.82	113.5	+9.9
1877	134.2	4.90	128.5	+5.7
1882	156.5	5.05	145.5	+11.0
1887	162.1	5.09	164.7	-2.6
1892	181.1	5.20	186.4	-4.6
1897	202.1	5.31	211.0	-8.9
1902	227.1	5.43	238.9	-11.8
1907	267.8	5.59	270.4	-2.6

To find the compound growth equation, $\ln(1 + r) = c$ or $1 + r = 1.0251$. The average annual growth rate is about 2.5%.

The values of Q-Q' over the fifty year period show that the approximation is comfortably within 10% of the true value for all years except 1857 and within 5% from 1887.

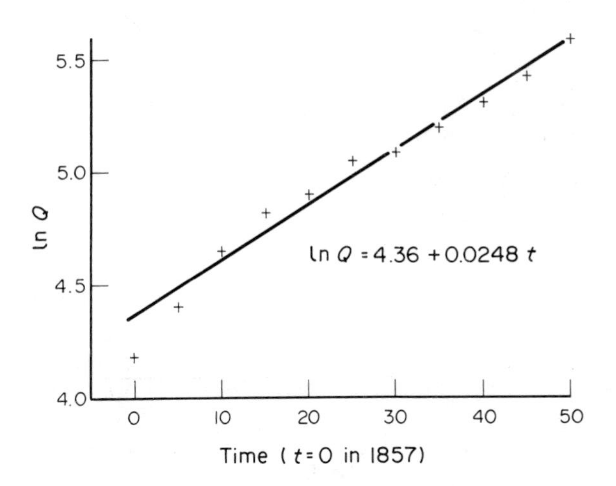

Figure 1.4

There are practical limitations to these exponential growth rate equations. If a government advisor had used this equation in 1907 to predict the UK output in 1957, it would have been 935 million tons per annum.

The actual output in that year was 207.6 million tons.

Exponential decay

The best known example of exponential decay, or negative exponential growth, is radioactive decay. At any instant the decay rate is directly proportional to the number of radioactive atoms present. If N is the number of radioactive atoms present at time t, then:

$$\frac{dN}{dt} = -\lambda N \qquad (1.19)$$

where λ is called the decay constant. The general solution is $N = N_0 e^{-\lambda t}$, where $N_0 = N$ when t=0. The half-life of a radioactive element is the length of time needed for the number of radioactive atoms to be halved.

Example 1.3 If the decay constant for carbon 14 is 0.0001201, what is its half-life?

$$1 = 2e^{-0.0001201t}$$

$$t = \frac{\ln 2}{0.0001201} = 5771 \text{ years}$$

Radioisotope tables usually list the half-lives rather than decay constants.

The practical limits to growth

It is clear from the results of Example 1.2 that exponential growth cannot continue without other factors intervening. In the UK there was more than enough coal available for all needs at the start of the twentieth century. The greatest output was achieved in 1913 with 287.4 million tons and over one million men employed in the mines. Scarcity of the resource was not a problem then, nor is it today. But with all three fossil fuels the total quantity of the resource is limited to a finite value. The later stages in production could be represented by a 'limited growth' model, in which the rate of increase is given by

$$\frac{dQ}{dt} = c(Q_f - Q) \qquad (1.20)$$

where Q_f is the finite limit for the particular resource.

To give an initial exponential rise in production followed by a symmetrical decline, the rate of growth over the whole life cycle is given by the product of the exponential growth model and the limited growth model, so that

$$\frac{dQ}{dt} = cQ(Q_f - Q) \qquad (1.21)$$

This is called the logistic equation. It is commonly used in the life sciences to examine the growth and regulation of populations[2,3]. The solution is

$$\frac{Q}{Q_f} = \frac{1}{1 + ae^{-cQ_f t}} \qquad (1.22)$$

where **a** and **c** are constants, and is shown in Figure 1.5

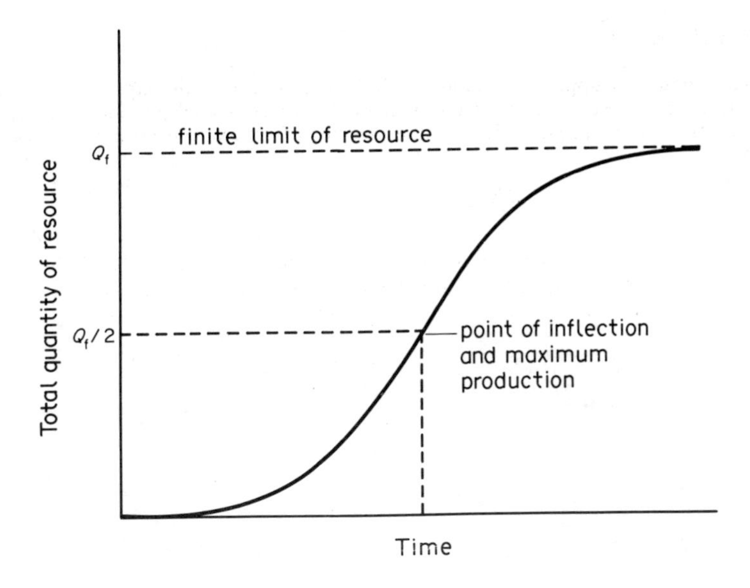

Figure 1.5 The logistic curve

The total possible life cycle of the three main fossil fuels and also of many minerals has been studied extensively. There is widespread agreement that the production of each particular fuel or mineral follows through the same sequence of events, although on completely different time scales. The first stage consists of the initial discovery, accompanied by the establishment of some mining and production facilities. This, in turn, encourages greater exploration and more reserves are discovered, as shown in Figure 1.6.

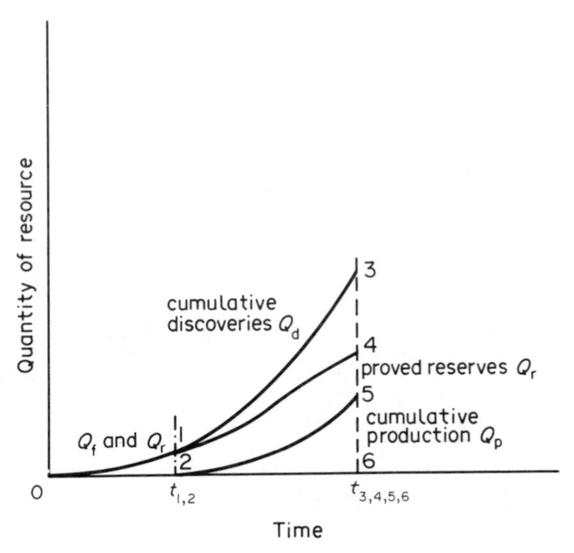

Figure 1.6

At point 1 the cumulative reserves and discoveries are equal and production is about to start, point 2. Market demand creates the incentives for more discoveries and production. Both the curves of cumulative discoveries, Q_d, and cumulative production, Q_p, start to increase exponentially. The difference between Q_d and Q_p, shown in Figure 1.6 gives the proved reserves at that time, Q_r. This can be plotted on the same base and gives the point 4. The value (4-6) is equal to (3-5). At this point the annual production level is still less than the annual increase in discovered reserves. The future for both producer and consumer is highly satisfactory. This trend continues until the new reserves found during the year are less than the annual production and at about the same time the level of production reaches its maximum value. From this point the logistic curves follow the paths shown in Figure 1.7.

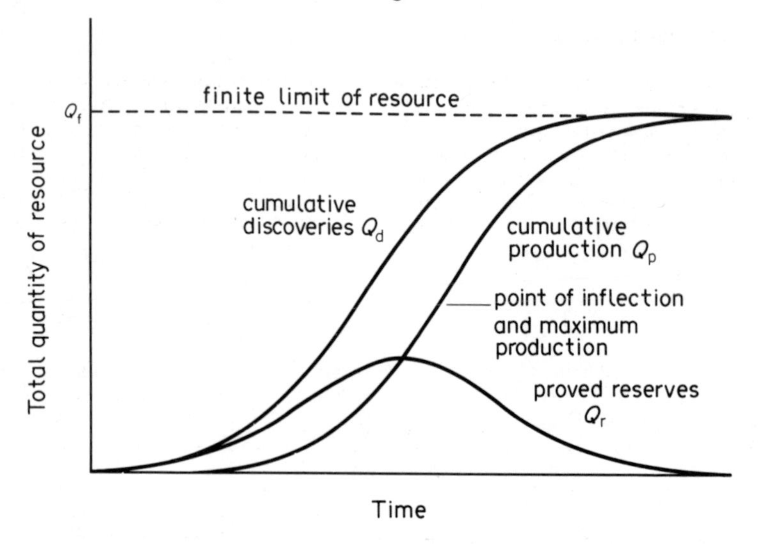

Figure 1.7

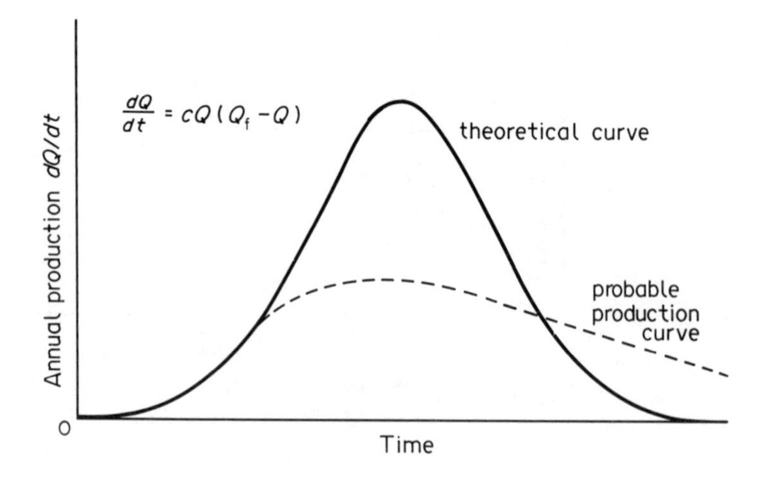

Figure 1.8

In practice the price of the resource would rise as production levels started to fall. This would probably reduce demand to well below that predicted by the logistic curves and lengthen the probable lifetime of the resource.

This is shown in Figure 1.8, where the theoretical production curve, $\frac{dQ}{dt}$, is shown with a more probable production prediction.

The maximum output per annum will be at least 50 times smaller than Q_f for all the fossil fuels. Other factors influencing annual production and consumption rates will be discussed in the next chapter.

The Earth's Energy Flows

There is a continuous flow of energy through the earth's atmosphere and surface. By far the greatest energy source is the sun. The other two sources, heat from the earth's interior and tidal energy causes by the gravitational forces of the earth-moon-sun system, are almost negligible in comparison. A world energy flow diagram is shown in Figure 1.9. The main flow paths were originally suggested by Hubbert[4] in 1962, who subsequently quantified all of them in 1971[5]. Some of these values have been revised to include later work, but it must be appreciated that all the values are approximate.

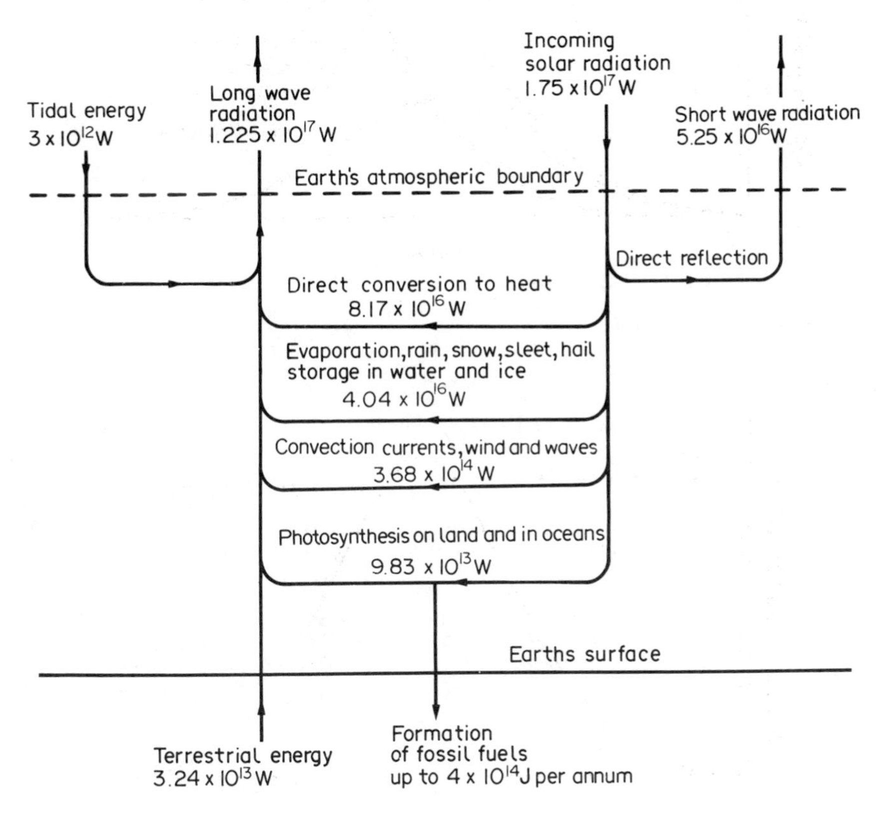

Figure 1.9 The earth's energy flows

Energy from Solar Radiation

The rate at which the incoming solar radiation is intercepted at the edge of the earth's atmosphere is $1373 Wm^{-2}$ with a probable error of 1-2%[6]. This is known as the 'solar constant', defined as the energy received by a unit surface perpendicular to the solar beam at the earth's mean distance from the sun. During the year the solar constant can vary by $\pm 3.4\%$, partly due to variations in the earth-sun distance. The total solar power intercepted by the earth is given by $1373\pi r^2$, where r is the mean radius of the earth, 6.371×10^6m. This gives a figure of 1.75×10^{17} watts or an annual total of 5.52×10^{24}J. For comparison, the world's total commercial energy consumption in 1982 was 2.86×10^{20}J[7], or some twenty thousand times smaller than the solar energy intercepted by the earth.

Energy from the Earth's Interior

Heat flow by conduction through the earth's solid crust has been estimated to be approximately $0.063 Wm^{-2}$[5]. Although there are a few areas surrounding active volcanoes and hot springs where the heat flow can be very much greater than this value, these sources are estimated to contribute at a rate of perhaps 1% of the conduction rate. The conduction rate is given by $0.063 \, 4\pi r^2$, where $4\pi r^2$ is the earth's surface area. This gives a figure of 3.21×10^{13} watts for conduction and a total of approximately 3.24×10^{13} watts for the terrestrial energy flow.

Energy from the Tides

Tidal energy has been estimated from studies of the rates of change of the periods of rotation of the earth and moon to be 3×10^{12} watts[5].

The Energy Flow System

Some 30% (5.25×10^{16}W) of the incoming solar radiation is directly reflected and scattered back into space as short-wave radiation. The earth's atmosphere, the oceans and the land masses absorb about 47% (8.17×10^{16}W). This is converted directly into heat at the ambient surface temperature and is reradiated as long-wave radiation. The hydrological cycle uses about 23% (4.04×10^{16}W). As water evaporates and changes into water vapour, heat is absorbed. This heat is released when the water is precipitated. The water vapour is often convected high up in the atmosphere. When precipitation occurs, water from these high levels provides all the potential energy available in lakes and the kinetic energy of rivers. Most is this energy is dissipated as low temperature heat as it descends to sea level. A very small proportion of the incoming solar radiation, about 0.21% (3.68×10^{14}W), creates the movements in the earth's atmosphere which cause the winds and ocean waves and currents. These are also dissipated as low temperature heat.

The amount of organic matter synthesised annually on land and in the oceans by photosynthesis is equivalent to 3.1×10^{21}J of stored energy[8]. This represents less than 0.06% (9.83×10^{13}W) of the incoming solar radiation, but is nevertheless about ten times the world's total commercial energy consumption in 1982. Photosynthesis is the conversion of solar energy into the fixed energy of carbohydrates through the reaction:

$$\text{Solar Energy} + CO_2 + H_2O \rightarrow \text{Carbohydrates} + O_2$$

About 0.5% of this photosynthetically fixed carbon is consumed annually as food[9] and up to 1.5% is directly consumed throughout the world as fuel wood. During the past

600 million years the fossil fuels, coal, oil and natural gas, were formed from a very small fraction of the synthesised organic matter. This matter is laid down each year in places such as swamps and peat bogs where its rate of decay is greatly retarded. The annual rate of formation can be calculated by dividing the estimated total fossil fuel resources, approximately 2×10^{24} J, by the total formation period. This gives a figure of up to 4×10^{14} J. In these calculations the various potential energy resources from nuclear fuels have been omitted.

Summary

There are many different forms of energy. The first law of thermodynamics states that energy is never 'consumed', it can be transformed from one form to another. The second law describes the thermodynamic limitations to any energy transformation. In addition, all real energy transformations suffer from energy degradation.

In a finite world unlimited growth is impossible. This applies equally to population, wealth or the production and consumption of any fossil fuel.

References

1. O'Callaghan, P.W. and Probert, S.D. Energy and Economics, Applied Energy 8, 227-243, 1981.

2. De Sapio, R. Calculus for the Life Sciences, W.H. Freeman and Company, San Francisco, 1978.

3. Ross, M.H. and Williams R.H. The potential for fuel conservation, Technology Review 79,4, 1977.

4. Hubbert, M. King, Energy Resources: A Report to the Committee on Natural Resources, National Academy of Sciences - National Research Council. Publ. 1000-D, Washington D.C., 1962.

5. Hubbert, M.K., The Energy Resources of the Earth, Scientific American, 225, 61-70 September 1971.

6. Frolich, C., Contemporary measures of the solar constant, in The Solar Output and its Variation, Colarado Associated University Press, Boulder, 1977.

7. B.P. Statistical Review of World Energy 1982, The British Petroleum Company plc, London, May 1983.

8. Hall, D.O., in Solar Energy: A UK Assessment, U.K. Section, International Solar Energy Society, London, p266, 1976.

9. Hall, D.O. and Rao, K.K., Photosynthesis, Edward Arnold, London, 1976.

Further Reading

A more rigorous treatment of the concepts of heat and work, together with the first and second laws of thermodynamics will be found in many undergraduate texts on thermodynamics. In the U.K. the Open University second level course, Thermofluid Mechanics and Energy, T233, is particularly recommended. The classical work on the application of logistic curves to energy resources is M. King Hubbert's Energy Resources, reference 4 above, and also in Resources and Man, a report for the US

National Academy of Sciences, W.H. Freeman and Company, 1969. A non-mathematical approach to the topics discussed in this chapter will be found in Gerald Foley, The Energy Question, Second Edition, Penguin Books Ltd., 1981.

Exercises

1. A body of mass 30kg is initially at rest on a ledge 20m above ground level. If it falls to the ground what is the change in the potential energy of the body?
   Choose one answer: (a) 5886J (b) 600J (c) -5.886kJ (d) -600J (e) none of these is correct.

2. A body of mass 5kg has a velocity of $300 ms^{-1}$. What is its kinetic energy?
   Choose one answer: (a) 225kJ (b) 450kJ (c) 1500J (d) -225kJ (e) none of these is correct.

3. A system has an initial internal energy, E_1, of 20kJ. After a net work input to the system of 100kJ the final internal energy, E_2, is 60kJ. What is the net heat transfer to the system?
   Choose one answer: (a) +40kJ (b) +60kJ (c) -40kJ (d) -60kJ.

4. Which of the following statements are correct according to the second law of thermodynamics?
   (a) A machine could be built to provide work by exchanging heat with only one very large heat source such as the sea, although it would be very expensive to build.
   (b) Heat cannot be converted completely into useful work.
   (c) Heat cannot pass spontaneously from a lower temperature body to a body at a higher temperature.
   (d) For the purposes of transforming heat into useful work it is better to have one unit of heat at 200°C rather than two units of heat at 100°C. The surroundings are at 50°C.
   (e) For the purposes of transforming heat into useful work it is better to have one unit of heat at 200°C rather than two units of heat at 100°C. The surroundings are at 0°C.

5. The amount of heat supplied for transformation into work from a high temperature source at 1000K is 200MJ. The low temperature sink is at 300K. How much of the heat supplied can be classified as 'available' heat under these conditions?
   Choose one answer: (a) 60MJ (b) 70MJ (c) 100MJ (d) 140MJ

6. In a small hydro-electric scheme water is held in a reservoir 200 metres above the power station. The conversion efficiency from potential energy to electrical energy is 82%. If the mass flow rate of the water through the power station is $25 kgs^{-1}$, what is the electrical power output?
   Choose one of the following: (a) 4.10kW (b) 5.00kW (c) 40.22kW (d) 49.05kW (e) none of these is correct.

7. The consumption of coal in a certain country was 10 million tonnes per annum last year. The energy forecasts suggest a compound growth of 10% per annum over the next ten years. What will the annual coal consumption be at the end of ten years?
   Choose one answer: (a) 20 million tonnes (b) 23.6 million tonnes (c) 25.94 million tonnes (d) 28.53 million tonnes.

8. A gas-fired heating boiler has a conversion efficiency of 72%. It is used to heat a building at a temperature of 22°C in surroundings at 0°C. What is its second law efficiency?
Choose one answer: (a) 5.14% (b) 5.8% (c) 51.4% (d) 72%.

Answers:

1 (c) 2 (a) 3 (d) 4 (b),(c),(d) 5 (d) 6 (c) 7 (c) 8 (a).

2. DEVELOPMENT OF ENERGY SOURCES

Early History

Life and energy flow from the sun[1]. The use of solar and other forms of energy has been the most important single factor influencing man's activities since the dawn of civilization. Improvements in physical comfort, food supplies and in many types of economic activity have depended upon an increasing supply of energy. The American energy analyst, M. King Hubbert, in a report to the Committee on Natural Resources in 1962[2] had no doubts about this when he stated

> ... it follows that the history of human culture must also be a history of man's increasing ability to control and manipulate energy.

Without the use of energy the quality of life can scarcely rise above the bare survival level of 'primative man', some 1,000,000 years ago in East Africa. He probably consumed only the energy equivalent of the food he could eat, just over 8MJ (about 2000 kilocalories) per capita daily, as he had not discovered fire[2]. This level was also used as the first of six levels of man's daily energy consumption per capita identified by Earl Cook[3]. With the second level, 'hunting man', consumption increased to about 20MJ (5000 kilocalories) daily some 100,000 years ago in Europe. He had learned to use some tools and burned wood for cooking and heating. Primitive 'agricultural man', about 5000 BC, had begun to use animal power and to grow crops. He consumed up to 50 MJ (12000 kilocalories) daily. Even today there are millions of people who live close to this level. They rely almost entirely on the direct and indirect use of solar energy for all their needs, including shelter.

From about 5000 BC, early cities with populations up to 10,000 became established in the Eastern Mediterranean countries, surrounded and supported by intensive agriculture[2]. At this time there was also evidence of the ability to mine metals and to forge them into tools and weapons using the heat energy from burning wood and charcoal.

Meanwhile, other uses of energy were developing in China. Their scientists and technologists, who could perhaps lay claim to the title of 'early scientific man' were probably the first people to use natural gas on an industrial scale. It was used to evaporate salt during the Qin (221 to 207 BC) and early Han (202 BC to 9 AD) dynasties[4]. At that time they were also producing metals and ceramics by methods which clearly required energy inputs substantially greater than those of the simpler agricultural activities outlined by Cook[3]. For example, thousands of full size pottery figures of men and horses with their associated bronze weapons have been found in the Imperial Tomb of Qin Shi Huang in Shaanxi Province. These show that the chief components of the swords and arrowheads were bronze and tin, with a small amount of rare metals, and that the surfaces had been treated with chromium.

The use of the principle of causing a vacuum in a vessel full of steam by suddenly cooling it was also described in writings attributed to that period and Needham comments[5]

> ... it is astonishing that a manifestation of that power which became available in seventeenth century Europe for raising water and for so many other purposes should first have been described as far back as the 2nd century in China.

Some ancient Chinese writings record the use of 'burning mirrors' for igniting tinder from the sun's rays[6]. Bronze mirrors are recorded as far back as the seventh century BC and their use to provide fire appears to have been fairly common in some religious ceremonies during the Han dynasty (202 BC - 220 AD). The Greeks and Romans were also developing burning mirrors at this time, with the alleged attack by Archimedes upon the Roman fleet at Syracuse in the year 214 BC being the best known early example. Many other applications of various systems for focusing the sun's rays with combinations of mirrors or lenses have appeared in the literature since then[1].

Deforestation, regarded today as one of the major problems facing the world, was also a problem in ancient Greece. By the fifth century BC, many parts of Greece had been stripped of trees and in the fourth century BC laws were passed in Athens forbidding the use of olive wood for making charcoal and banning fuel wood exports[7].

The fundamental principles governing the use of solar energy in architecture were also developed then, probably as a response to the fuel wood shortages. Socrates (470 - 399 BC) is believed to have been the earliest philosopher to point out that in south facing houses

> ... the rays of the sun penetrate into the porticos during the winter, but in the summer the sun's path is directly over our heads and above the roofs, so that there is shade. If, therefore, this is the best arrangement, we should build the south side higher, to catch the winter sun, and the north side lower to exclude the cold winds ...

It is probable that the first applications of water power appeared during the next few hundred years, as a watermill was described in a Greek poem during the first century BC[8]. The horizontal shafted nora or noria wheel was the simplest type and had flat paddles which were dipped into a flowing river. Its use was confined to raising water. The small vertical shafted Greek or Norse mill used water flowing down a steep wooden shute to strike the blades. These were originally flat but were sometimes curved to improve their efficiency in energy conversion. The height of the river was between one and two metres above the blades. The later, more efficient, Roman mill was first described by Vitruvius[9]. Its horizontal shaft was carried in two bearings and it turned the millstones by means of a right angled gearwheel, probably the first use of power transmission through gearing. The millstones usually made five revolutions to one of the water wheel. The waterwheel was placed in a specially shaped channel, so that its lower parts were completely submerged. This was known as an undershot waterwheel. One wheel installed near Naples was described as having a diameter of 1.85 m, a width of 0.3 m and an estimated output of just over 2 kW. It was capable of grinding 150 kg of corn per hour[8]. This was about 40 times more than the output of a slave.

Problems of unemployment caused by new technology were not unknown to the Romans. Vespasian (69 - 79 AD) is believed to have refused to sanction the building of a water-driven hoist 'lest the poor have no work'[8].

The Barbegal mill, built by the Romans in the south of France towards the end of the fourth century, had two parallel sets of eight overshot wheels, the water striking the upper section of each wheel in turn. It was installed on a slope of some 30° to the horizontal and the total head of water was 18.6 metres. Each wheel had a diameter of 2.2 m and a width of 0.7 m. Although this seemed to be a much larger system than the single wheel described above, its maximum corn output was some 320 kg per hour, just over twice the output of the single wheel. This is believed to be the largest mill for the next twelve centuries.

The earliest reference to a water mill in England mentions a mill near Dover in 762 AD, but by the time of the Domesday Book in 1086, 5624 water mills are listed in some 3000 communities south of the Trent and Severn[10]. The Romans had used waterwheels from boats anchored in flowing rivers during the seige of Rome in 537 but the tidal mill with solid foundations did not appear until the twelfth century. The same period shows that the principles of energy saving or conservation were understood and practised in China. Needham[11] quotes an example of a substantial improvement in the first law efficiency of an energy conversion system. The simple flat pan oil lamp which had been widely used for lighting in many ancient civilisations, held the oil and the wick. Because much oil was being lost by evaporation without efficient combustion, saucer-shaped cruses were made with a reservoir underneath to hold cold water. Needham recorded that he saw some of these 'economic lamps' in the Chunking Museum and that the twelfth century Chinese Lu Yu stated

> 'The flame of an ordinary lamp as it burns quickly dries up the oil, but these lamps are different for they save half the oil ....'

Wind-driven water pumps for irrigation may have first appeared in Persia from about the fifth century. These early Persian designs used cloth sails and had a vertical axis. The windmill became established in Western Europe from the twelfth century. The traditional horizontal axis tower mill for grinding corn, having its sails supported by a large tower rather than by a single post, appeared in several parts of Europe at the beginning of the fourteenth century[1].

When it was appreciated that the water mill could supply power which could be used for other applications in addition to corn grinding, new levels of technological progress were gradually achieved. Tanning mills were recorded in 1217, paint mills in 1361 and sawmills in 1376[8]. By 1400 'advanced agricultural man', the fourth level in Cook's analysis, was making use of several of the renewable energy resources, including the developments in water and wind power described above and wood. He was also using coal for heating, first known to the early Greeks and Romans[12], and animals for transport. His estimated daily consumption was about 110 MJ (26000 kilocalories) daily, just over twice as much energy as 'primitive agriculatural man' was using in 5000 BC. This represented an annual compound growth rate of per capita energy consumption over the 6400 year period of about 0.0125%, or just over one hundredth of one per cent.

Humphrey and Stanislaw[13], noted that the use of coal in the UK is believed to date from the twelfth century, but for several centuries it was considered to be technicaly and environmentally inferior to wood. They drew attention to the concern about environmental pollution expressed in a proclamation made by Edward II in 1307 forbidding the limeburners of Southwark to burn coal because

> ... an intolerable smell diffuses itself throughout the neighbouring places, and the air is greatly affected ...

During the sixteenth century the shortages of fuel wood which had been experienced in Greece almost two thousand years earlier became serious in England. A compound annual growth rate of 1.46% in the price of firewood from 1500 to 1642 was matched for a while by a slower price rise in coal, but when the price of coal started to fall again from about 1630, it replaced wood in the household market[14]. This was an early example of a fuel being in short supply, its price rising and the eventual substitution of an alternative cheaper energy source.

By the beginning of the sixteenth century it could be argued that comparatively little progress had been made in basic engineering applications since the first century AD. Water pumps and watermills were only slightly better than those described by the early Greeks and Romans. The books on architecture written by Vitrivius[9] were still being used in Italy. But from this point scientific and technological skills started to develop rapidly[13]. Scientific societies were founded almost simultaneously in England, Germany, Italy and France during the seventeenth century. The largest water pumping system built since the Roman mills at Barbegal was completed at the Palace of Versailles in 1682. About 4.5 million litres were raised daily to an aquaduct 153m above the river, giving a useful work output of some 80 kW[8].

But in south-east England there was a major fuel resource crisis. The iron industry relied on wood to make charcoal for smelting the iron and to a lesser extent on water power to drive bellows to maintain continuous blasts of air to raise the furnace temperature. Although a limited amount of coal was available in shallow mines or from surface outcrops, it could not be used to make iron with the technology available up to 1700. The shortage of wood caused a serious decline in the industry and pig iron had to be imported[15].

However, one of the practical scientific applications at the end of the seventeenth century was the use of the suction pump[13]. This probably helped Thomas Savery in the development of the first water pump to make use of steam power. Savery's 'fire' engine appeared just before 1700 and was powered by coal.

From the Industrial Revolution to the Twentieth Century (1700-1900)

Savery's engine had no piston. Water was pumped partly by direct steam pressure and partly by vacuum. A somewhat similar concept using solar energy to provide the heat, rather than coal, was subsequently described by Belidor[1] and is shown in Figure 2.1.

During the next few years two problems which had been holding back economic and industrial developments in the United Kingdom were solved. In 1709 Abraham Darby discovered how to smelt iron using coke derived from coal. This was followed by Thomas Newcomen's invention of the first steam engine to have a piston and cylinder in 1712. The piston was connected by a rod and chain to the arched end of an overhead beam and the steam was condensed at atmospheric pressure. These engines were known as 'beam' engines. Based on the amount of water raised by burning a known amount of coal, the overall efficiency of energy conversion of the first Newcomen beam engine was about 0.5%[16]. The performance of steam engines could also be measured by the proportion of the total heat of the steam supplied to the engine which was converted into 'indicated' work. Indicated work was calculated from the mean pressure in the cylinder and the volume of steam passing through it. This gave a theoretical 'indicated' thermal efficiency which was always greater than the overall efficiency because of the heat losses in the boiler, steam pipes and transmission.

Between 1730 and 1765 John Smeaton improved the overall efficiency of the Newcomen engine to about 1%, but this was still a very poor performance. The major breakthrough was achieved by James Watt in 1765. He improved the efficiency of the

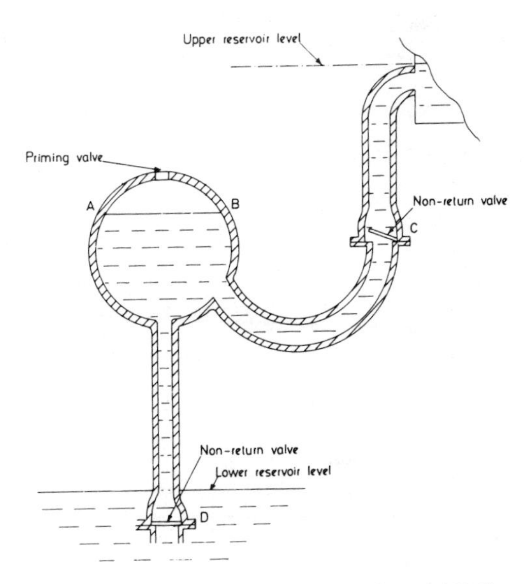

Figure 2.1. Belidor's solar pump (circa 1740)

The pump is primed by filling the spherical dome to the level AB. During the day solar radiation heats the dome, causing the air to expand and force the water through the non-return valve C to reach the upper reservoir. On cooling, either artifically or at night, the internal air pressure falls below atmospheric, drawing water into the pump from the lower reservoir through the non-return valve D.

steam engine by a factor of three or four by condensing the steam in a separate water-cooled condenser[16]. Subsequently he altered the vertical motion of the beam engine to the first example of a modern rotary engine, fitted with a connecting rod and crankshaft. The improvements in overall steam engine and steam turbine system efficiency which followed Watt are shown in Figure 2.2.

A third problem, the continuing shortage of wood, was described by Deane[19] as the major bottleneck in the British economy in 1750. This had already resulted in an increasing demand for coal for heating, especially in London. It meant that the coal mines had to be deeper as there was insufficient coal to meet demand from the surface outcrops. More water collected in the deeper mines and it became harder to pump out because of the increased pumping pressure. The early Watt engines enabled coal mining to be firmly established as steam powered pumps could be used and coal output increased rapidly from this point.

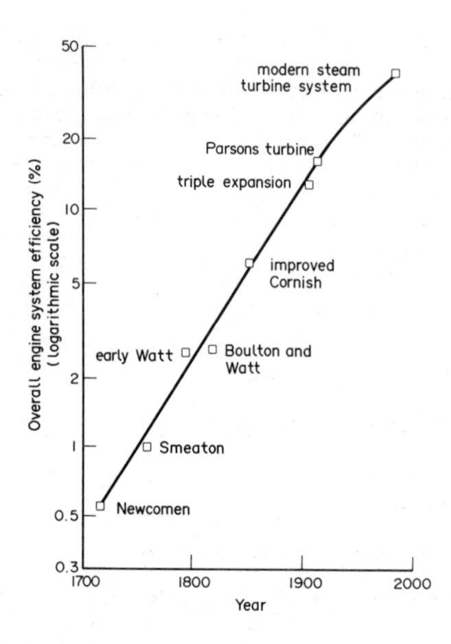

Figure 2.2 Steam engine and steam turbine efficiencies

(Data derived from references 16,17 and 18)

The nineteenth century saw the United Kingdom transform from a purely agricultural community to the first great industrial community in the world[16]. In 1800 the population was about 16 million and primary energy consumption was about 11 million tonnes of coal. But coal could now replace water and wind power in all types of mill. Ships could be driven by steam power instead of sails. On land the steam locomotive and the railway network would revolutionise transport. The railways removed another obstacle to industrial progress, the problem of moving the coal from the mines to the factories. In parallel with the growth in industrial activity come improvement in agriculture. For example in the period up to 1900, the number of men required to harvest wheat at a rate of 10 acres a day fell from 126 (entirely by hand) to 21 as, by 1900, the steam thresher had been added to the reaper and binder[20]. Further improvements in equipment meant that more land could be brought into production. The development of the economy in the United Kingdom during the nineteenth century is summarised in Table 2.1.

Table 2.1

Development of the United Kingdom Economy 1800-1900

	1800	1900	Ratio 1900/1800
Population (millions)	16	41.5	2.6
Coal consumption (million of tonnes)	11	165	15
Coal consumption per capita	0.68	4.00	5.9
Industrial engine capacity (horsepower)	4000	10 000 000	2500
Index of output per capita	100	985	9.85

(Data derived from references 14, 16 and 20)

The early nineteenth century also saw the start of the gas industry and the discovery of electricity. Simple drilling techniques to reach brine wells for salt production were also common, but drilling for oil had never been attempted well drilled by Edwin L. Drake, near the banks of Oil Creek in Pennsylvania, struck oil at a depth of 21.2 metres on the 27th August 1859, although he had been looking for salt on behalf of the chemical industry. This has been widely regarded as the world's first oil well, but there were many earlier recorded examples of the use of oil in various forms throughout history, several of which are discussed in the next chapter. Many other significant developments in energy applications followed Drake's well in the second half of the century. Some of these are listed in Table 2.2.

Table 2.2

Some significant energy developments 1800-1900

Gas lighting (in Paris)	1801
Electricity discovered (Michael Faraday)	1831
Oil discovered in Pennysylvania (Edwin Drake)	1859
Internal combustion engine (Lenoir)	1860
Solar powered steam engine (Mouchot)	1860
Solar powered refrigeration (in Paris)	1878
Electric lighting (Swan, UK)	1878
Hydro-electric power (Northumberland, UK)	1880
Electrical power station (Steam-powered, London and Brighton, UK, followed by New York)	1882
Commercial solar water heaters (USA)	1891
Electricity generation from town refuse (Halifax, UK)	1893
Modern motor car (Panhard)	1894
Wind-generated electricity (Denmark)	1895

But as early as 1872 there were warnings that coal would eventually be exhausted. The Swedish American engineer John Ericsson[21] wrote

> The time will come when Europe must stop her mills for the want of coal. Upper Egypt, then, with her never-ceasing sunpower, will invite the European manufacturer to ... erect his mills ... along the sides of the Nile.

It is important to appreciate that the increase in coal consumption was not directly caused by the use of coal in power applications. The increase followed because it became possible to get the coal out to the surface with the steam-powered pumps.

The Twentieth Century

Although coal accounted for nearly 95% of the commercial energy in the world at the beginning of the century, several leading scientists and engineers in the United Kingdom were aware of its limitations. They recognised that this period could only be regarded as 'a first step in the mastery of energy' and were looking into the future. The radiochemist, Professor Frederick Soddy[22], wrote in 1912

> This is the Age of Energy, or rather this is the beginning of the ages of energy, the Age of the Energy of Coal. That this still is the age of the energy of coal is unfortunately only too true, and the whole earth is rendered the filthier thereby. Moreover, the age will last just so long as the coal supply lasts, and after that the last state of the race will be worse than the first, unless it has learned better.

Soddy had worked with Rutherford in the earliest experiments on the nature of radioactivity and was able to state that this work put an entirely different complexion on the question as to how long the energy resources of the world may be expected to last. Correctly predicting nuclear power he said

> ... there exists in matter, associated ... with the smallest particles capable of separate existence ... sufficient potential energy to supply the uttermost ambitions of the race for cosmical epochs of time.

Professor A.H. Gibson, a Civil Engineer, also drew attention to the concern which had been expressed about the rapid depletion of the coal resources of the United Kingdom and other actively industrial countries[23]. Pessimists were saying that these resources might only last a few hundred years and optimists many thousands. Gibson reviewed all the possible known alternative energy sources and concluded that even when the fossil fuels are exhausted, ample supplies of renewable energy would remain for all the conceivable activities of the human race. As many existing oil fields in the United States were already becoming depleted, Gibson concluded that oil would be exhausted long before coal. He drew attention to Schaler's prediction[24] that the supply of oil would not sensibly outlast the present century, but felt this outlook was probably too pessimistic. Dr. Dugald Clerk[16], a Mechanical Engineer, gave figures for world coal reserves broadly similar to the upper limits suggested in the 1980s, and drew attention to the need for energy conservation. He also pointed out the advantages of using the exhaust heat from power stations to provide heat for space-heating and industrial process applications. This concept was adopted many years later in both district heating and combined heat and power (CHP) schemes. Clerk also believed that oil was probably not important, particularly as a report from the United States Geological Survey considered that at the rate of increase in consumption experienced at that time, oil would be exhausted by 1935.

Nevertheless the strategic significance of oil as a transport fuel was appreciated by Winston Churchill, the First Lord of the Admiralty in 1913, who ordered the Navy to change to oil-firing. Shortly afterwards the Admiralty bought a 51% share in the Anglo-Persian Oil Company, later the British Petroleum Company[25].

Professor Frederick Soddy also spelt out the limitations of economic theories when these were applied to energy[26]. In 1921 he stated that

> ... the economist is peculiarly liable to mistake for laws of nature the laws of human nature ... The principles and ethics of human law and convention must not run counter to those of thermodynamics ...

He was also the first to draw a distinction between the continuous 'revenue' of energy received from the sun and the use of the 'capital' energy stored in the fossil fuels. He predicted that the period of prosperity through which Great Britain had passed was destined to be short lived. This prosperity depended on the ability to substitute the capital energy of fuel for labour, but civilisation ultimately relied on the revenue of solar energy for its food supply.

During the second half of the nineteenth century the United States and the UK were the world's leading commercial energy producers and consumers. In 1900 the United States (254 million tonnes) and the UK (230 million tonnes) produced just over 60% of the total world coal production of 780 tonnes. The importance of coal in creating national wealth and the key role of the coal miners led to conflict. Conditions were very poor for the miners compared with some other industries. At times they could withdraw their labour in an attempt to obtain higher wages and better conditions. In the United States, Charles Pope[27] wrote

> The year 1902 has added an awful chapter to the history of our need of a new source of heat and power, by the wide suffering and impoverishment of the people of the eastern United States in consequence of the Pennysylvania coal strikes....

and in England, a few years later, Gibson[23] commented

> The coal strikes of recent years afford a striking lesson of the suffering and dislocation of the social system ... by even a temporary shortage of the coal supplies.

Since 1900 the relative importance of coal has declined as first oil and then natural gas took increasing shares as shown in Table 2.3.

Table 2.3

World Primary Energy Consumption Structure 1900-1980
% of total

	1900	1920	1940	1960	1980
Coal	94.2	86.7	74.6	52.1	29.1
Oil	3.8	9.5	17.9	31.2	43.5
Natural Gas	1.5	1.9	4.6	14.6	18.9
Hydroelectricity	0.5	2.0	2.9	2.1	6.1
Nuclear electricity	-	-	-	-	3.1

Coal includes lignite. The contribution from nuclear electricity was less than 1% in 1960. Additions for 1920, 1960 and 1980 vary due to rounding-off errors. Data from references 28 and 29.

The development of the energy patterns of the United Stated and the United Kingdom in 1900 appeared to be fairly equal. Their commercial energy consumptions per capita, at about 4 tonnes of coal equivalent per capita, were practically identical. The major difference was that indigenous oil and natural gas had already started to contribute 3% and 2% respectively to the total energy consumption in the United States. Coal dominated consumption in the UK for the next fifty years, with a share of over 90% of total consumption. The earliest figures[30] for the consumption in the USSR date from 1913. Coal and oil contributed 48% and 30% to their total consumption of just under 50 million tonnes of coal equivalent with peat and wood providing the balance.

Growth in world energy consumption continued steadily from 1900 to 1950 as shown in Figure 2.3.

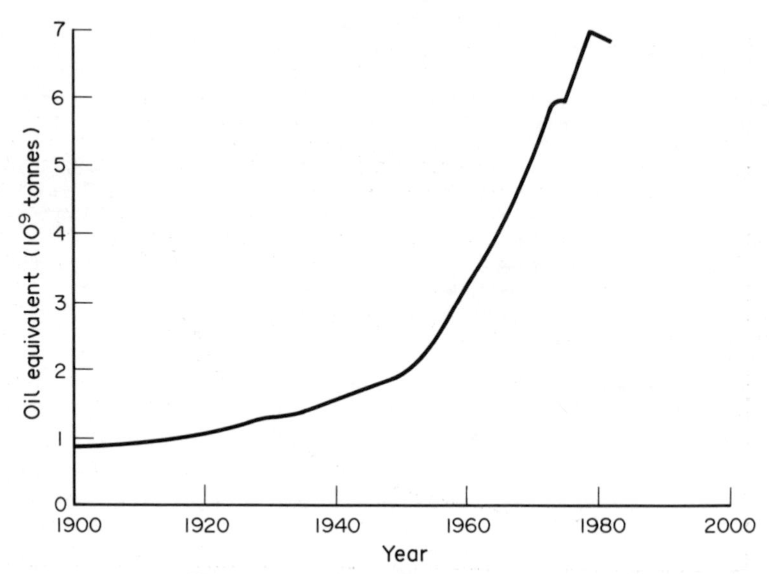

Figure 2.3 World energy consumption

This period contained two world wars, 1914-1918 and 1939-1946, separated by the major world depression of the 1930s, when it was impossible to make meaningful assessments of long term trends. Estimates of world population have been available for a much longer period[31,32,33] and the growth pattern is shown in Figure 2.4. From 1900 the rate of population growth appears to be increasing exponentially.

By 1950 the United States had firmly established itself as the world's major energy consumer and producer. Consumption had risen to about 1360 m.t.c.e., a fourfold increase since 1900, with a doubling of per capita energy consumption. The United Kingdom had a much smaller relative increase in energy consumption, about 40% over the 50 year period to 229.5 m.t.c.e. The USSR was emerging as the second most powerful energy producer, and showing signs of following the growth pattern of the United States, having a consumption of 311.2 m.t.c.e., with natural gas at 5% and oil at 17% of their total consumption .

This was also a time for reassessment. The American, Palmer C. Putman, who had been responsible for the development of the world's largest windmill in the 1940s, was asked by his Government to examine trends in energy consumption and population growth. His report[31], prepared from data available in 1950, was published in 1951. Although the extent of both oil and natural gas resources were not fully apparent at that time, he restated the earlier warnings of Soddy, Gibson and Clerk on the implications of energy resource limitations, and predicted the 'energy crisis' of 1973 within two years!

> If present trends continue in population growth, in demand per capita for energy output as measured at the point of end use and in the preferences for fluid fuels and for electrical energy ... then the real unit costs of energy from the harder-to-win and poorer-rank coal will cause a strong demand for new sources of energy sooner than many had realized - a demand which may become insistent by 1975 AD.

The trends in population growth mentioned above all suggested that the world's population would rise to 3900 million ± 200 million by the year 2000, shown as the 'expert' projection trend in Figure 2.4.

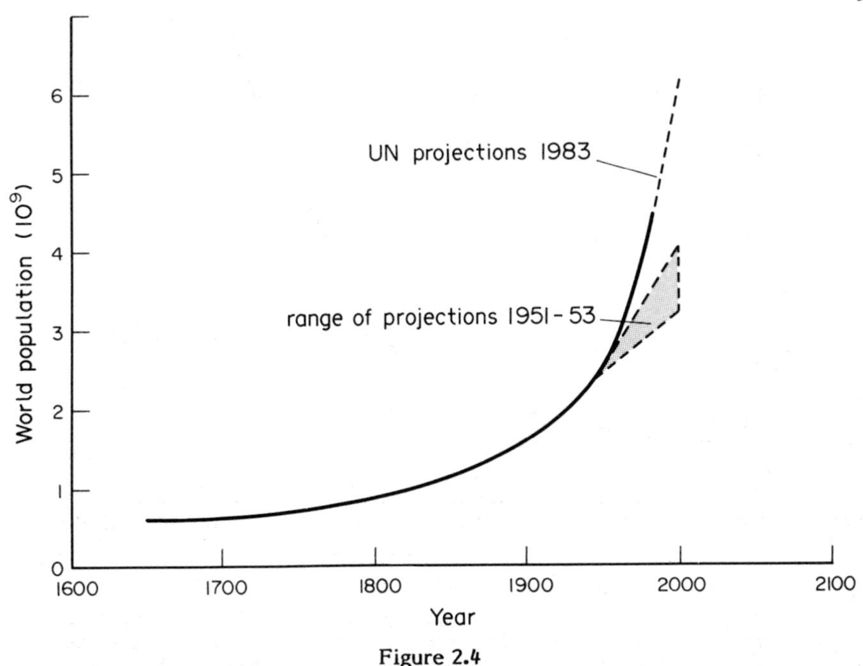

Figure 2.4

Putman concluded that when the cumulative estimates of the net economically available reserves were compared with the maximum plausible demands, the peak production of fossil fuels may not be far away. Over the period from 1900 to 1970, growth in energy consumption in the United States averaged 2.93% per annum and in the UK 0.94% per annum. The growth rate in the USSR from 1950 to 1970 was 6.7% per annum. Figure 2.3 shows that the total world energy consumption doubled between 1950 and 1965.

Growth, Crisis and Reassessment, 1965-1982

The exponential growth rates in energy consumption and in population were not the only problems facing the world in the 1960s. Exponential growth rates were also being experienced in the consumption of many minerals and in some forms of environmental pollution. This led to the first attempts to develop a theoretical global computer model of the interaction between these and other factors, including agricultural and industrial production. The work arose from discussions held by a group of some thirty individuals from ten countries who first met in Rome in April 1968 under the leadership of Dr. Aurelio Peccei. The group subsequently grew to about seventy people from twenty-five countries and was known as The Club of Rome. Their report[34], published in 1972, predicted that the world could not continue to support the rates of economic growth and population growth which were then being experienced. Although the report's assumptions of continuing exponential growth were criticised, it stimulated further studies and anticipated the corrective trends which started to appear some ten years after its publication. These are discussed in Chapter 10.

Oil production reached its maximum point in the United States in 1970 in spite of Government incentives over the previous twelve years through oil import controls. Two years later warnings of the vunerability of Western Europe, the United States and Japan to incipient oil shortages were made by Tiratsoo[28]. He anticipated that the United States would need to increase its exports of goods and services to world markets during the early 1980s to support its trade deficit in fuels and that this could profoundly affect the pattern of world trade. He also pointed out that oil prices were likely to double by 1975 following the results of 1971-72 negotiations between the producing states and the international oil companies. His perceptive analysis of the world energy situation cannot be bettered today. Many months before the 1973 'energy crisis' he wrote

> It seems clear therefore, that, the problem the developed countries of the world now have to face is not so much that of a forthcoming oil shortage, but rather the fact that the era of cheap oil on which our modern civilisation has largely been built has now come to an end.

The events which occurred in October 1973 have been subsequently regarded as the beginning of the 'energy crisis'. Briefly, during October 1973 the Syrians and the Egyptians attacked Israel through the Golan Heights and across the Suez Canel. A few days later the Arab oil producers raised their oil prices by 70% and decided to cut production by 5% each month. In December, a second very much larger price rise was announced by Iran. This resulted in an overall quadrupling of the price charged by the oil producers during a year. The second major event occured in 1979 when supplies from Iran, then the world's second largest oil exporter, were terminated during the revolution. Oil prices doubled during the year and subsequently rose a further 30% on the doubled price during 1980. The effects of these latest price rises resulted in a severe world recession. Higher prices reduced demand. Economic growth was curtailed and unemployment rose. These events are reflected in the discontinuities shown from 1973-5 and from 1979 in Figure 2.3.

The sequence of events from 1965 to 1982 can now be studied in greater detail. Tables 2.4 to 2.8 inclusive examine the trends in energy consumption and oil production in selected countries and regions. Figure 2.5 shows how the world's primary energy consumption was shared between the major energy sources in 1965 and 1982, with 1973, 1950 and 1930 included for comparison.

The events of 1973 made the Industrial Economies re-assess their use and production of energy. They had all benefited when oil prices were relatively low in the 1950s and 1960s. The percentage of their oil consumption relative to their primary energy consumption had risen from less than 40% in 1960 to 53.3% in 1973. Table 2.4 shows that over the period from 1965-73 they were consuming over 70% of the total world oil production, but that their own production had fallen from meeting nearly half their demand, to about a third. They failed to appreciate the political and economic consequences of relying so much on oil imports. Table 2.4 can be used to calculate their oil import dependency. In 1965 oil represented 43.8% of their primary energy consumption and they produced 45.3% of this oil. In other words 54.7% of their oil was imported. Their oil import dependency was therefore 0.547 x 43.8%, or 24%. By 1973 the corresponding figure had risen to 0.661 x 53.3% or just over 35% of their primary energy consumption.

Figures 2.3 and 2.5 taken with Tables 2.5 and 2.6 show the exponential rise in consumption experienced in all sectors from 1965 to 1973. The growth in world primary energy consumption of 4.6% per annum shown in Table 2.5 is equivalent to a doubling period of only 15 years. Figure 2.5 shows that doubling took place in the 15 years from 1950 to 1965 and that in the eight years from 1965 to 1973 the increase in world energy consumption was equivalent to the total consumption in 1950.

Table 2.4

+ Relative importance of oil in the Industrial Economies

	1965	1967	1972	1973	1977	1982
Percentage of their oil consumption relative to their Primary Energy Consumption	43.8	45.8	52.2	53.3	52.0	45.2
Percentage of their oil consumption relative to the total World Oil Consumption	72.0	71.3	70.3	69.9	65.2	57.5
Percentage of their oil production relative to total World Oil Production	31.9	31.1	25.1	23.1	20.7	26.4
Percentage of their oil production relative to their total oil consumption	45.3	44.9	36.2	33.9	32.6	44.8

\+ author's derived data

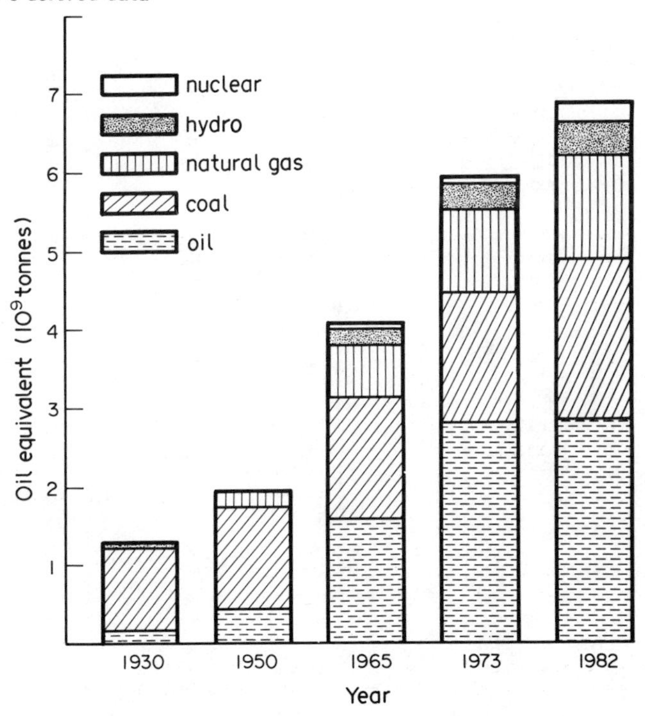

Figure 2.5 World primary energy consumption

Table 2.5

World Primary Energy Consumption

Region/Country	Million tonnes Oil Equivalent						+ Yearly Change %		
	1965	1967	1972	1973	1977	1982	1965-73	1973-77	1977-82
USA	1340.0	1460.8	1767.8	1822.7	1846.8	1728.4	+3.9	+0.3	-1.3
Canada	126.2	140.9	179.7	190.9	213.0	207.4	+5.3	+2.8	-0.5
Western Europe	852.0	904.4	1168.9	1240.9	1259.5	1217.3	+4.8	+0.4	-0.7
Japan	153.6	196.7	310.8	347.7	348.1	340.2	+10.8	+0.03	-0.5
Australasia	44.8	49.9	63.9	66.9	80.3	89.5	+5.1	+4.8	+2.2
+ Industrial Economies Total	2516.6	2752.7	3491.1	3668.8	3747.7	3582.8	+4.8	+0.5	-0.9
USSR	648.7	723.8	836.5	874.1	1060.6	1242.0	+3.8	+5.0	+3.2
Eastern Europe	266.5	277.0	353.9	357.2	423.6	455.4	+3.7	+4.4	+1.5
+ Centrally Planned Economies Total	915.2	1000.8	1190.4	1231.3	1484.2	1697.4	+3.8	+4.8	+3.4
China	230.7	180.8	334.9	362.5	459.6	522.1	+5.8	+6.1	+2.6
+ Rest of World	458.5	517.2	614.3	660.6	799.1	1032.0	+4.7	+4.9	+5.2
World	4121.0	4451.5	5630.7	5923.2	6490.6	6834.3	+4.6	+2.3	+1.0

+ Indicates figures derived by the author

Table 2.6

Oil Consumption

Region/Country	Million tonnes						+ Yearly Change %		
	1965	1967	1972	1973	1977	1982	1965–73	1973–77	1977–82
USA	548.9	595.8	775.8	818.0	865.9	703.0	+5.1	+1.4	-4.1
Canada	55.0	61.7	79.3	83.7	85.6	73.0	+5.4	+0.6	-3.1
Western Europe	388.5	458.1	701.8	748.9	697.3	601.2	+8.5	-1.8	-2.9
Japan	87.9	122.9	234.4	269.1	260.4	207.0	+15.0	-0.8	-4.5
Australasia	20.8	23.5	31.7	34.8	38.0	35.5	+6.6	+2.2	-1.4
+ Industrial Economies Total	1101.1	1262.0	1823.0	1954.5	1947.2	1619.7	+7.4	-0.1	-3.6
World Total	1529.2	1769.3	2592.4	2798.0	2986.9	2818.8	+7.8	+1.6	-1.2

Notes for Tables 2.4, 2.5, 2.6, 2.7 and 2.8

Basic data is taken from the B.P. statistical review of the world oil industry 1975 and the B.P. Statistical Review of World Energy 1982. The B.P. classifications of Western Europe, Eastern Europe and Australasia, as defined below, are used. The two other groups, the Industrial Economies and the Centrally Planned Economies, are as defined by the author.

Western Europe consists of the members of the European Economic Community (Belgium, Denmark, France, Greece, Republic of Ireland, Italy, Luxembourg, Netherlands, United Kingdom, West Germany) plus Austria, Finland, Iceland, Norway, Portugal, Spain, Sweden, Switzerland, Turkey, with Yugoslavia, Cyprus, Gibralter, Malta and the Canary Islands.

Eastern Europe consists of Albania, Bulgeria, Czechoslovakia, East Germany, Hungary, Poland, Romania.

Australasia consists of Australia, Papua New Guinea, New Zealand, South West Pacific Islands.

The Industrial Economies consist of the USA, Canada, Western Europe, Japan and Australasia. Data for this group will be within 1% of that for the OECD (Organisation for Economic Development and Cooperation) countries, as the relative energy consumption of the non-OECD countries (Cyprus, Gibralter, Malta, the Canary Islands, Papua New Guinea and the South West Pacific Islands) is very small.

The Centrally Planned Economies consist of the USSR and Eastern Europe.

+ Indicates figures derived by the author.

Their re-assessment had to achieve two objectives - the reduction of their dependence on oil and an improvement in the end use efficiency of all types of energy conversion systems. By 1982 most of the countries in the Industrial Economies had prepared energy policies which emphasized energy conservation and the substitution of other fuels for oil where possible. They had increased their efforts to develop their indigenous energy resources and to carry out new research, development and demonstration projects with alternative energy sources. The 'Coconuc' policy - coal, conservation and nuclear energy - was widely mentioned in Western Europe. World energy consumption reached a peak in 1979* followed by a decline of about $\frac{1}{2}$% per annum to 1982. This is reflected in the overall figures for the five year period from 1977 to 1982, which disguise a two year growth period followed by a more rapid decline than the rates indicated in Tables 2.5 and 2.6 By 1982 the oil import dependency of the Industrial Economies had fallen to 0.522 x 45.2% or 25% of their primary energy consumption. Table 2.4 also shows how their share of world oil production fell until the production from the North Sea became significant in 1982. The increasing need for oil to aid development in the rest of the world was an additional reason for the fall in the share of world oil consumption taken by the Industrial Economies. Table 2.7 shows that their percentage share of world primary energy consumption fell from the 62% level, where it had been from 1965 to 1973, to 52.4% by 1982.

Table 2.7

+ Primary Energy Consumption of the three main groups expressed as a percentage of World Primary Energy Consumption

	Year					
Group	1965	1967	1972	1973	1977	1982
Industrial Economies	61.1	61.8	62.0	61.9	57.7	52.4
Centrally Planned Economies	22.2	22.5	21.1	20.8	22.9	24.8
China and Rest of World	16.7	15.7	16.9	17.3	19.4	22.8

+ Author's derived data

The Centrally Planned Economies have been able to continue their economic progress without disruption to their energy growth pattern because of the enormous energy resources within the USSR. Their energy consumption actually increased over the four year period from 1973-77 compared with the previous eight year period and was still increasing in 1982. Table 2.7 also shows that their share of the world primary energy consumption had started to rise by 1982 because of the overall reduction in the consumption of the Industrial Economies. For the rest of the world, Table 2.7 shows their energy consumption rising from 16.7% to 22.8% of the world total. Some reasons for this are discussed later. Trends in the relative importance of oil in four countries are examined in Table 2.8. Japan has by far the greatest dependence on oil, but nevertheless had managed a significant reduction from the 77.4% value in 1973 to just over 60% by 1982. The United States took rather longer to react than the UK, but both have achieved reductions to 40%. China has a rapidly developing economy and this is reflected in the rising share of oil in their energy consumption over the whole 17 year period.

* The year 1979 can be examined in greater detail in the first exercise at the end of this Chapter

Table 2.8

+ Relative importance of oil in selected countries. Percentage of their oil consumption relative to their primary energy consumption

	1965	1967	1972	1973	1977	1982
USA	41.0	40.8	43.9	44.9	46.9	40.7
UK	38.8	43.7	51.3	50.4	43.4	39.1
Japan	57.2	62.5	75.4	77.4	74.8	60.8
China	4.5	6.1	12.9	14.8	17.8	15.8

\+ Author's derived data.

Energy Consumption Balance and Population

The pattern of energy consumption in six countries during 1982 is shown in Figure 2.6

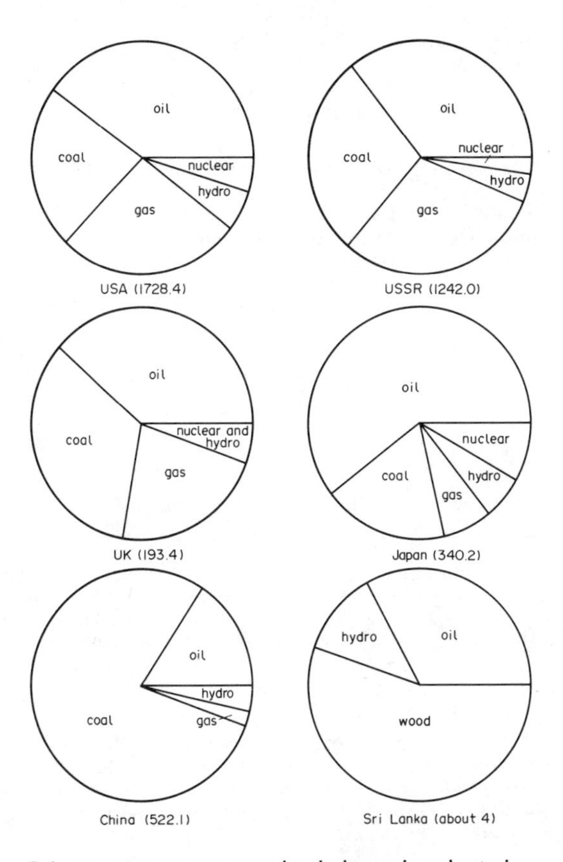

Figure 2.6 Primary energy consumption balance in selected countries (1982) (million tonnes of oil equivalent)

This pattern is influenced by many factors, including energy resources, climate, economic structure and technical development. Three of the six countries, the USSR, the UK and China, are self-sufficient in energy resources. The two largest energy consumers in the world, the United States and USSR show very similar patterns of energy consumption. Both had very substantial shares of indigenous coal, oil and natural gas. By the end of the 1970s, the United States had used a much greater proportion of its energy resources than the USSR. The consumption pattern in the UK is also very similar to that of the United States and the USSR. This is due to the developments of its North Sea gas and oil resources since 1965 as well as its traditional coal base. Most of Japan's energy is imported and the country is heavily dependent on oil imports. China and Sri Lanka are both regarded by the World Bank as low income developing countries. China has a well-developed coal industry which dominates their supply and consumption pattern. Oil production is comparatively modest, but new contracts for joint exploration with Western countries were agreed in 1983. Sri Lanka still relies mainly on wood as the dominant primary energy input, but oil has by far the greatest share in end-use energy. The hydro-electricity share will increase considerably during the 1980s.

When the size of population is also considered, the distribution of energy per capita shows great inequalities between the different countries as shown in Table 2.9.

Table 2.9

Relative energy consumption and population, selected countries (1981-2)

	Population (millions)	Percentage of world population +	Percentage of world energy consumption +	Energy consumption per capita (t.o.e.) +
USA	229.8	5.16	25.28	7.52
USSR	268.0	6.01	18.17	4.63
Japan	117.6	2.64	4.97	2.89
UK	56.0	1.25	2.83	3.45
China	991.3	22.25	7.64	0.53
Sri Lanka	15.0	0.33	0.03	0.13

Sources: The World Bank[32] and the British Petroleum Company[29]. Energy consumption is based on commercial energy figures. Energy consumption per capita is given in tonnes of oil equivalent (t.o.e.)

+ author's derived data.

The United States, with only one-twentieth of world's population consumes a quarter of the world's commercial energy. The United States and the USSR consume nearly half the world's commercial energy with just over one-tenth of the world's population shared between them. China, with nearly a quarter of the world's population consumes about one fifteenth of the world's energy.

Energy consumption per capita averaged over the world is 1.53 t.o.e. The average person in the United States consumes five times this average. In Sri Lanka and in many other developing countries the figure is less than a tenth of the average.

Energy Production and Consumption in the United Kingdom.

Since 1950 the United Kingdom has seen several major and quite unpredictable changes in its pattern of energy consumption and production. This is shown in Figure 2.7.

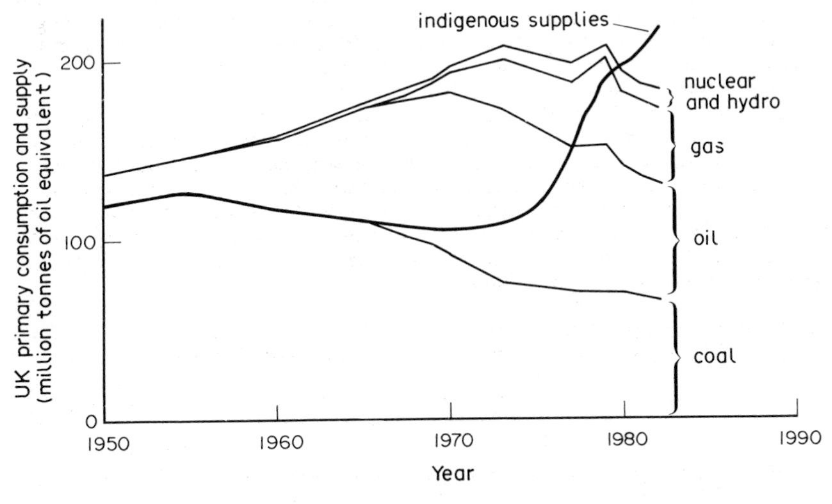

Figure 2.7

Although coal output rose slightly from 1950 to 1956, its contribution to total energy consumption fell during the decade from 90% to just under 75%, being replaced by low cost oil. The first reactor in the UK civil nuclear programme commenced generating electricity for the National Grid in 1956. At that time it was hoped that nuclear power would be able to make a significant contribution to UK energy demands. However, it took a further twenty years before it first supplied 4% of total primary energy consumption. In 1964 the UK was still a two-fuel energy economy, but on the 26th December drilling for oil and gas commenced in the UK sector of the North Sea. Gas was found about 70 km east of the Humber by the British Petroleum Company about ten months later. This was the West Sole gasfield, the first to supply gas by pipeline to the mainland at Easington and then into the National Gas Grid system. This was followed by other finds fairly close to West Sole in the southern North Sea, the Leman field being the largest. Since then further discoveries have been made in the northern sector between the Shetland Islands and also in Morecambe Bay in the Irish Sea. By 1970 Natural Gas was contributing over 5% to the UK total energy consumption. In November 1970 the British Petroleum Company found oil in what is now known as the Forties field and development during the next decade was dramatic. Oil first came ashore in 1975 and by 1976 the Department of Energy reported[35] that the North Sea oil and gas industries had already created 100,000 new jobs. Towards the end of 1977 North Sea production was half total consumption and by 1980 the UK was producing more oil than was consumed. The Primary Fuel Production figures from 1973 to 1982 are given in Table 2.10 and the corresponding consumption figures in Table 2.11. The main features are as follows:

Coal

The decline experienced in the industry from the mid 1950s was halted in 1974 when the rise in oil prices made coal competitive again. This relative cost advantage over oil was increased by the 1979-80 price rises in oil. By the end of 1983 coal stocks had reached record levels, as output was continuing to exceed the reduced demand partly caused by the general recession.

Oil

Rising production levels and falling consumption levels resulted in a very healthy export market by the end of 1982.

Table 2.10

UK Primary Fuel Production
(million tonnes of oil equivalent)

	Coal	Oil	Natural Gas	Nuclear Electricity	Hydro Electricity	Total
1973	77.6	0.4	25.3	5.9	1.2	110.4
1974	65.0	0.4	30.6	7.1	1.3	104.4
1975	75.7	1.6	31.8	6.4	1.2	116.7
1976	72.8	12.1	33.7	7.6	1.1	127.3
1977	71.9	38.2	35.3	8.4	1.2	155.0
1978	72.7	54.0	33.9	7.9	1.2	169.7
1979	72.0	77.9	33.7	8.1	1.3	193.0
1980	76.5	80.5	32.1	7.8	1.2	198.1
1981	75.0	89.5	31.9	8.1	1.3	205.8
1982	73.4	103.4	32.4	9.4	1.4	220.0

Table 2.11

UK Primary Fuel Inland Energy Consumption+
(million tonnes of oil equivalent)

	Coal	Oil	Natural Gas	Nuclear Electricity	Hydro Electricity	Total
1973	78.2	96.6	26.0	5.9	1.2	207.9
1974	69.3	89.7	31.2	7.1	1.3	198.6
1975	70.6	80.3	32.6	6.4	1.2	191.1
1976	71.8	78.9	34.6	7.6	1.1	194.0
1977	72.2	80.3	36.9	8.4	1.2	199.0
1978	70.5	82.0	38.3	7.9	1.2	199.9
1979	76.2	81.8	41.8	8.1	1.3	209.2
1980	71.1	71.4	41.8	7.8	1.2	193.3
1981	69.6	65.2	42.4	8.1	1.3	186.6
1982	65.1	65.4	42.2	9.4	1.4	183.5

+ Note this does not include 'non-energy' uses, such as feedstocks for petrochemical plants or lubricants. The figures in other tables for UK consumption include these non-energy uses. In 1982 these amounted to about 10 m.t.o.e.

Gas

Demand and supply have been in balance throughout the period. The difference between UK production and consumption arises from the Norwegian sector of the Frigg field in the North Sea. The UK could be self-sufficient in gas today, but as the Norwegians were unable to find an economical method for landing their gas in Norway, it is bought by the UK under a long-term contract. Long-term negotiations were already underway in 1983 to purchase Norwegian gas from their Sleipner field in the 1990s[36].

Nuclear Electricity

After the delays experienced with Advanced Gas Cooler Reactor Programme during the 1970s, additional capacity was coming on stream during 1983. The output for the first 9 months of 1983 indicated that the final annual production would exceed 10 m.t.o.e. for the first time[37].

The use of fuel in the UK during 1980 is illustrated by the simplified energy flow diagram shown in Figure 2.8. The production of useful or 'end use' energy consumes 30.1% of the total primary energy input. This waste is larger than the consumption in any other single end use sector. Energy flow diagrams are useful for examining overall trends.

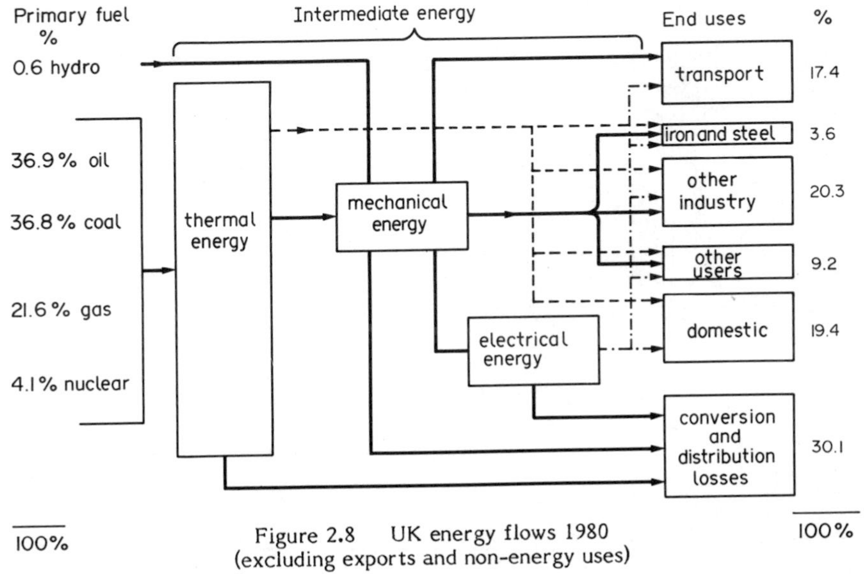

Figure 2.8 UK energy flows 1980
(excluding exports and non-energy uses)

By 1980 the UK was in an excellent position. Energy production was exceeding consumption, with the net production curve rising, as shown in Figure 2.7. Scope for further economies in the end-use of energy were clearly evident from Figure 2.8. The future prospects for UK energy will be discussed in Chapter 10.

Gross National Product and Energy Consumption

Economists try to assess the national wealth of a country by examining the costs of all the various activities within the country. The basic inputs to any particular activity can include labour, capital, raw materials and energy. Inputs can come from both home and foreign resources and can include scientific and technical skills. The outputs are in the form of manufactured goods of all types, including food and services. These outputs are demanded by government or groups such as consumers or foreign countries. The current market value of all these goods and services is known as the Gross Domestic Product (GDP). When the earnings from any foreign investments are added to the GDP the result is the Gross National Product (GNP). There are very large differences in the GNP of different countries, just as there are very large differences in their energy consumption patterns or size of population.

Data used in any GNP - energy consumption analysis must be regarded with caution. Statistical methods and practices used in different countries for calculating GNP often vary quite widely and problems arise from variations in their exchange rates with the US dollar. Full details of the methods used to derive GNP data have been given by the World Bank[32]. They point out that the relative purchasing power of national currencies is not accurately measured by the use of official exchange rates. Factors converting the energy value of one fuel to another also vary. Only commercial energy is considered in official statistics. The energy consumption figure from many developing countries will appear to be very low, as they ignore the contribution from fuel wood which could account for as much as 90% of their total energy consumption. A futher complication is that figures for energy consumption may subsequently be readjusted when more complete data becomes available.

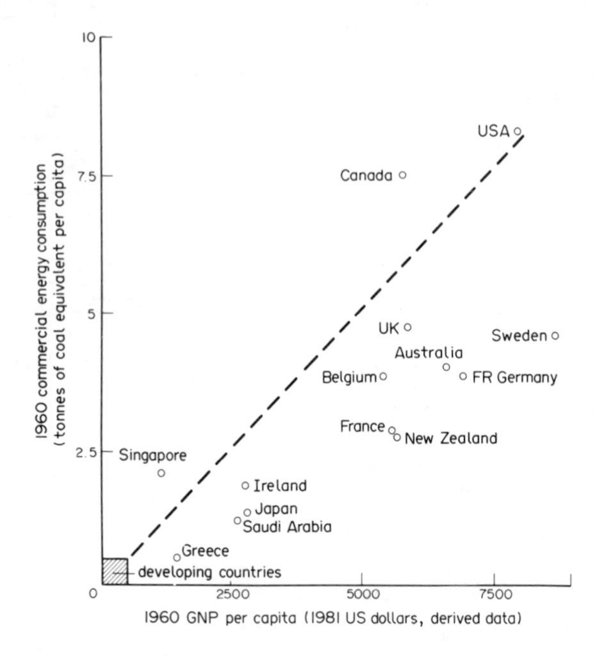

Figure 2.9

During the 1960s it was found that there was an approximate correlation between the energy consumption per capita and the GNP per capita of those countries for whom reliable data could be obtained. Figure 2.9 shows these trends for a number of selected countries in 1960. It would have been interesting to include some countries from the Centrally Planned Economies, but their data was not available on a comparable basis. The GNP per capita has been derived form 1981 data using growth rates over the period given by the World Bank[32]. The derived GNP data has been used rather than historic data so that the trends over the period from 1960 to 1981 can be studied in greater detail in Figure 2.10. An examination of historic GNP - energy consumption data for the 1960s gives a broadly similar picture, but may show the United States further ahead in GNP[3].

Figure 2.9 shows that greater wealth, as measured by GDP, is associated with higher per capita energy consumption. Some countries have less energy intensive industries than others or may be more efficient in their use of energy. This would place them further towards the bottom right-hand corner. The UK, for example, had approximately the same GDP as France, but partly through having a colder climate and poorly insulated housing and partly through having more heavy industry at that time the energy consumption per capita in the UK was greater. Some of the developing countries included in the group shown in the bottom left hand corner are considered by the World Bank[32] to be closer to the Industrial Economies in their real incomes than the apparently wide differences shown by GNP per capita in US dollars for the reasons outlined in the previous section.

During the 1960s it could be argued that as a country increased its GNP, the additional wealth encouraged more growth in energy consumption. By the end of the 1970s considerable doubts had been expressed about the apparent direct relationship between per capita energy consumption and GNP. Long term energy forecasts using this relationship had been found to overestimate energy consumption.

The trends in per capita energy consumption and GNP of the countries studied in Figure 2.9 are shown in Figure 2.10. Over the twenty one year period each country increased both its GDP per capita and energy consumption per capita. The range of paths is so diverse that a simple direct relationship should no longer be used. Compared with countries in the Industrial Economies group or the lower income developing countries, the newly industrialising countries appear to need a greater amount of energy to produce a unit of GDP. Initially their path will tend to follow the Singapore trend, rather than the more energy conserving path of countries in Western Europe such as the UK, France or Germany. The spectacular growth rates in the economies of Saudi Arabia and Japan have been achieved with very much smaller energy consumptions per capita than Canada or the United States. Canada's economy grew at faster rate than that of the United States, but at the expense of a greatly increased per capita energy consumption. Five countries listed in the World Bank Report[32], Kuwait, Luxembourg, Switzerland, the United Arab Emirates and Qatar, had a GNP per capita greater than 15,000 US dollars in 1981.

The main use of an Energy Consumption per capita - GNP per capita diagram, such as Figure 2.10, is that an appreciation of the relative performance of several countries over a period of time can be obtained quickly. It cannot be used for detailed discussions, as the circumstances and factors influencing the performance of each country are so varied.

Summary

The history of man's use of energy shows that some of today's concepts were known to the ancient Chinese, Greek and Roman civilisations. The Industrial Revolution allowed the exploitation of coal to change a few of the agricultural-based economies into the early industrial economies of the nineteenth century. The modern industrial economies were developed during the era of cheap oil. This ended in the 1970s. Since then the world has been trying to readjust to a different pattern of energy consumption. The problem of providing suitable conditions for the economic growth of the developing countries in spite of greatly increased oil prices has still to be solved.

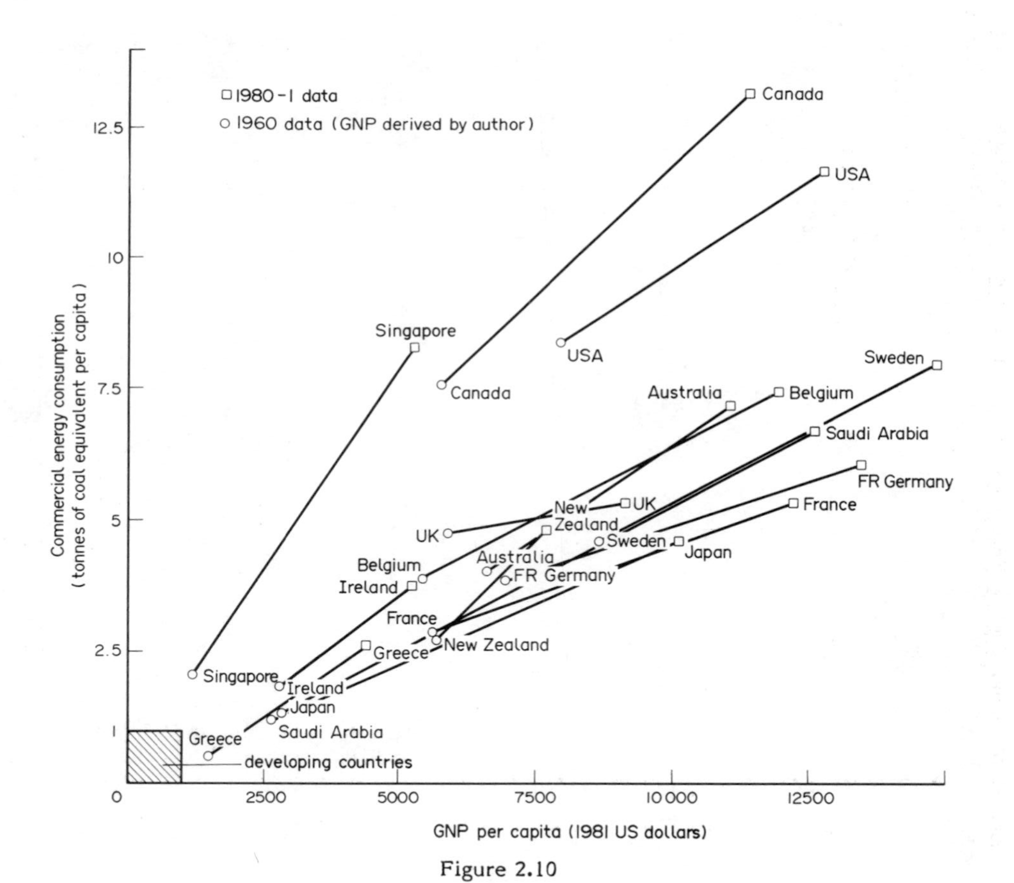

□ 1980-1 data
○ 1960 data (GNP derived by author)
Commercial energy consumption
(tonnes of coal equivalent per capita)
GNP per capita (1981 US dollars)
0
1
2.5
5
7.5
10
12.5
2500
5000
7500
10000
12500
Canada
USA
Singapore
Canada
USA
Sweden
Australia
Belgium
Saudi Arabia
FR Germany
New
Zealand
UK
France
UK
Sweden
Japan
Belgium
Australia
FR Germany
Ireland
France
Greece
New Zealand
Singapore
Ireland
Japan
Saudi Arabia
Greece
developing countries

Figure 2.10

References

1. McVeigh, J.C., Sun Power, Second Edition, Pergamon Press, 1983.

2. Hubbert M. King. Energy Resources. A report to the Committee on Natural Resources of the National Academy of Sciences - National Research Council. Publication 1000-D., National Research Council, Washington, D.C. 1962.

3. Cook, Earl, 'The Flow of Energy in an Industrial Society', Scientific American, 225, 135-144, September 1971.

4. Needham, Joseph, Science and Civilisation in China, Vol. IV, Part I, 66, Cambridge University Press, 1962.

5. Ibid, 70.

6. Ibid, 87-88.

7. Butti, K. and Perlin J., A Golden Thread, Cheshire Books, Palo Alto and Van Nostrand Reinhold, New York, 1980.

8. A History of Technology, Oxford University Press, II, 593-611, 1956.

9. Vitruvius. De architectura, written in the reign of Augustus (27BC-10AD), quoted in Ref. 8., above, p. 397.

10. Hodgen, M.T., Antiquity, 13, 261, 1939, quoted in Ref. 8, above, p. 611.

11. Needham, Joseph, Science and Civilisation in China, Vol IV, Part I, 78-9, Cambridge University Press, 1962.

12. A History of Technology, Oxford University Press, II, 369, 1956.

13. Ibid, III, 324, 1957.

14. Humphrey, W.S. and Stanislaw. J., Economic growth and energy consumption in the UK, 1700-1975, Energy Policy, 7, 29-42, 1979.

15. Nef. J.V. The Rise of the British Coal Industry, Vols I and II, George Routledge and Sons Ltd., London 1932 .

16. Clerk, Dugald., The World's supplies of fuel and motive power. Third Thomas Hawksley Lecture, Proc. 1 Mech. E., 89, 591-624, 1915.

17. A History of Technology, Oxford University Press, IV, 164, 1962.

18. Sassin, Wolfgang, Energy., Scientific American 243, 3, 119-132, 1980.

19. Deane, P., The First Industrial Revolution, Cambridge University Press, 1965.

20. Wright, S.J. Mechanical Engineering and Agriculture; Proc. 1. Mech. E., 1947, quoted in L.T.C. Rolt, The Mechanicals, W. Heinmann Ltd., London, 114-5, 1967.

21. Ericsson, J., Contributions to the Centennial Exhibition, John Ross, New York, 1876.

22. Soddy, Frederick, Matter and Energy, Williams and Norgate, London, 1912.

23. Gibson, A.H. Natural Sources of Energy, Cambridge University Press, 1913.

24. Ibid, 11.

25. Ezra, Derek (ed)., The Energy Debate, Benn Technical Books, Croydon, Surrey, 1983.

26. Soddy, Frederick, Cartesion Economics, Two lectures delivered in London University, November 1921; Henderson, London, 1924.

27. Pope, C.H., Solar-Heat - its practical applications, Boston, Massachusetts, 1903.

28. Tiratsoo, E.N., Oilfields of the World, Scientific Press Ltd., Beaconsfield, England, 1973.

29. B.P. Statistical Review of World Energy 1982, The British Petroleum Company plc, London, May 1983.

30. Elliott, Iain F., The Soviet Energy Balance, Praeger Publishers, Inc., New York, 1974.

31. Putnam, P.C. Energy in the Future, Macmillan & Co. Ltd., London, 1954.

32. The World Bank, Development Report, 1983.

33. Mahler, Halfdon, People, Scientific American 243, 3, 67-77, 1980.

34. Meadows, Dennis L. et. al., The Limits of Growth. A report for The Club of Rome's Project on the Predicament of Mankind. Potomac Associates, Inc., Washington 1972 and Pan Books Ltd., London 1974.

35. Development of the Oil and Gas Resources in the United Kingdom, Department of Energy, London, 1977.

36. Annual Report and Accounts, British Gas Corporation, 1982-83, London, HMSO, 1983.

37. Energy Trends, Department of Energy, HMSO, London, November 1983.

Further Reading

Many historical references to energy use and applications can be found in the Oxford University Press series A History of Technology, published in a number of volumes from 1956, and in Dr. Joseph Needham's Science and Civilisation in China, published by the Cambridge University Press. The history of solar energy is well covered by Ken Butti and John Perlin in A Golden Thread, published by Cheshire Books and Van Nostrand Reinhold, 1980. The Limits to Growth, the Club of Rome report has been published by Pan Books in a paperback edition. A well-balanced analysis of the issues discussed by the Club of Rome is given in Lord Ashby of Brandon's 21st Fawley Foundation Lecture, A Second Look at Doom, published by the University of Southampton, 1975.

The World Development Reports published for the World Bank by the Oxford University Press give very comprehensive data on GNP, population and commercial energy consumption. These reports include detailed assessments of general economic trends and issues. The British Petroleum Company's annual Statistical Review of World Energy also surveys the previous ten year's data, with a geographical survey of the main fossil fuel reserves and transport routes. The sources of UK statistical data are the Annual Abstract of Statistics and the UK Digest of Energy Statistics, published by HMSO London.

Exercises

1. The following Table gives data for the Industrial Economies in 1979:

Millions of tonnes of oil equivalent

	Oil Production	Oil Consumption	Primary Energy Consumption
USA	480.9	868.0	1918.6
Canada	83.8	90.1	221.8
Western Europe	116.5	732.4	1327.6
Japan	0.5	265.1	369.9
Australasia	21.9	38.1	86.1

   (a) What was the oil import dependency of the Industrial Economies in 1979?

   (b) What was the percentage yearly change in the primary energy consumption of the USA and Western Europe in the period from 1979 to 1982?

   (c) What factors could account for the values in (b) being substantially different from those in the period 1977-1982 given in Table 2.5?

2. Calculate the percentage yearly change in primary energy consumption for the period 1967-72 for the countries and regions listed in Table 2.5. Are there any significant differences between these results and the percentage yearly change figures given for 1965-73?

3. Plot a graph of the annual oil consumption of the United States since 1965, using Table 2.6 and the latest annual consumption figures. What is the percentage yearly change from 1982? Why has this changed from the -4.1% figure seen in the period from 1977-1982?

   (a) If the +5.1% figure for the yearly change of oil consumption in the United States between 1965 and 1973 had continued after 1973, when would the annual United States oil consumption have reached 1000 million tonnes?

   (b) If the -4.1% figure for the yearly change of oil consumption in the United States experienced between 1977 and 1982 were to continue, when would consumption be half the 1977 figure?

4. In 1981 the figures for GNP per capita for the United States, Japan and the United Kingdom were (in US dollars) 12,820; 10,080 and 9,100 respectively. During the period 1960 to 1981 their respective growth rates in GNP per capita were 2.3%, 6.3% and 2.1%. What were their figures for GNP per capita in 1960, also expressed in 1981 dollars?

5. In 1876 Thomas Hawksley, a former President of the Institution of Civil Engineers, discussed the growing population of England and Wales. He stated that, at the then rate of increase, the population of 24 million in 1876 would become 42 million in 1918 and, with the same rate of increase, 400 million at the end of the twenty first century, "..an obviously impossible population for so small a country".

(a) What was the annual percentage growth rate used by Hawksley between 1876 and 1918?

(b) How accurate was his figure of 400 million using the growth rate determined in (a)?

6. Examine the end-use energy in your school, college or workplace. What are the relative proportions of the different energy uses?

   Calculate the primary energy inputs necessary to achieve the end-use energy. How have these quantities changed over the past twenty years?

7. Distinguish between 'capital' energy and 'revenue' energy as defined by Professor Frederick Soddy. Why is this distinction important? Why was so little attention paid to Soddy's forecasts in 1921?

8. Dr. Joseph Needham has found evidence that Newton was not the first person to state what has been generally accepted as 'Newton's first law of motion'. Examine the history of the discovery of this law in its various forms.

Answers

1. (a) 32.6% (b) -3.4%, -2.84% (c) Mainly world recession from 1979.

2. +3.9%, +5.0, +5.3, +9.6, +5.1, +4.9, +2.9, +5.0, +3.5, +13.1, +3.5, +4.8 See China, Rest of World, Eastern Europe

3. (a) 1977, (b) 1994

4. 7952, 2794, 5888

5. (a) 1.341% (b) Correct figure is about 475 million, but his conclusions were still valid.

3. PETROLEUM

Introduction

Although the word petroleum is commonly considered to be a modern term, it dates from the sixteenth century and comes from the latin words petra, a rock, and oleum, oil. Today petroleum is used as a general term covering a wide range of hydrocarbons from natural gas, through crude oils of increasing viscosity to bitumen and solid paraffin waxes[1]. Most geologists believe that petroleum was derived from organic materials formed by the reactions of the remains of plants and marine animals buried in sedimentary rocks. The most significant periods for the geological formations in which petroleum has been discovered are in the early Tertiary period from 50 to 65 million years ago, the Cretaceous period from 65 to 136 million years ago and the Jurassic period from 136 to 190 million years ago. Sedimentary basins laid down in the early Tertiary period include parts of California and the United States Gulf Coast, Mexico and the Caribbean, Iran, Iraq and southern regions of the USSR. Basins laid down in the Cretaceous period include parts of Texas, Kuwait and Bahrain while the Jurassic period includes Saudi Arabia. The earliest oils have been found in the Precambrian rock formations and predate the formation of coal by hundreds of millions of years, long before the appearance of vegetation on the land masses[1].

As the marine residues of plants and animals settled in the bottom of the sea, the earlier layers were buried by successive layers above them. These earlier layers were compressed and heat was generated. This heat, combined with bacterial and chemical action, transformed the organic matter into various hydrocarbons. The sedimentary rocks, sandstone, limestone and shale, were also formed from the settlement of sands, silts and other marine forms. The sandstones are fairly porous and the volume of the porous spaces can be up to one third of the total volume. Initially these spaces were filled with water when the sedimentary basins were formed, but as gas and oil are less dense than water they tended to rise through the water until they became trapped under an impermeable or non-porous rock layer. If no trap is reached, the petroleum escapes to the surface and disperses. Not all the liquids and gases found in an oilfield have necessarily originated as petroleum. Some methane could have originated from coal, but both this methane and the petroleum gases and oils will have migrated slowly through the porous rocks to become trapped and form the oilfield.

A wide variety of geological traps have been identified. Many were caused by movements and folds in the earth's crust as mountain ranges were formed and regions at the bottom of the sea became dry land. Among the commonest are the fault, a fracture in the earth's crust which moved a non-porous layer vertically beside a porous layer, cutting it off; the anticline, rather similar to a semi-circular arch of impermeable rock above the porous layer; and the salt dome, caused by the upward movement of a massive salt flow which may have ruptured some sedimentary layers and caused others to arch. As the salt dome passed through any porous, oil-bearing sedimentary layer, oil could be trapped above it, or in faults at its sides[3,4]. An important type of trap which was not caused by the various structural deformations outlined above is the stratigraphic trap[3]. A rock layer may change laterally in composition or it may die out and reappear as a different type of rock. These changes can result in a lateral decrease in porosity and permeability in the same layer and the more porous sections may form a stratigraphic oil trap. A second type of stratigraphic trap can occur when a partially eroded permeable sandstone is covered by an impermeable rock layer. Some traps only contain oil, probably because the lower viscosity gas associated with it was able to migrate upwards through rocks which the oil was unable to pass through. Other traps may only contain gas. There are several possible explanations for this. The most likely is that with the higher temperatures and pressures associated with increasing depth, conditions suitable for oil formation did not exist below a depth of about 7000 metres and only gas could be formed. The majority of the world's oil fields cover only a few square kilometres, but some of the largest fields extend to several hundred square kilometres.

The hydrocarbons present in the crude oils can be classified into three main groups as follows[4]

(i) The paraffins, alkanes or saturated hydrocarbons based on the formula C_nH_{2n+2}. The value of n may vary from 1 to 70, with gases in the range $1<n<4$, liquids in the range $5<15$ and the remainder solids.
(ii) The naphthenes or cycloalkanes based on the formula C_nH_{2n}
(iii) The aromatic or benzenoid hydrocarbons based on the formula C_nH_{2n-6}.

Crude oils may also contain small proportions of sulphur, nitrogen and oxygen compounds and traces of some other elements including vanadium, nickel, arsenic and chlorine. A major problem is the presence of sulphur compounds which could cause unacceptable corrosion and pollution if they are not removed in the refinery. An oil containing more than 0.5% by weight of sulphur compounds is called 'sour' or 'high sulphur', while oil with less than 0.5% by weight is called 'sweet' or 'low sulphur'.

A further classification is into 'light' or 'heavy' crude oil. Heavier molecules do not take up more space in a liquid or solid, so that heavy crude oils with a greater number of larger molecules have a higher specific gravity than the lighter crudes. Specific gravity is the ratio of the density of the oil to that of water at standard conditions. The American Petroleum Institute (API) scale is widely used and its relation to specific gravity over the most common range for crude oils is given in Table 3.1. Most of the world's petroleum lies between 25 and 35° API.

Table 3.1

Specific Gravity at 60/60°F	API degrees
1.0000	10
0.9340	20
0.8762	30
0.7796	40
0.7389	50

History

The history of petroleum use goes back to the dawn of civilization[4]. Oil derived asphalt was obtained from hand-dug pits in the river valleys of Mesopotamia and used in building construction as flooring and waterproofing materials. Petroleum oils derived from seepages were used medicinally in ancient Egypt and later in the early Roman Empire, where boats were caulked with asphalt. Other uses for asphalt included jewel-setting and the preservation of mummies[4]. After the decline of the Roman Empire the technology reappeared in the Arab countries where the first distilling process was developed. The lighter oils obtained were used to fill porous pots which could be ignited by fuses and gunpowder. These early incendiary devices were reported in battles dating from the seventh century[4].

In Europe several sources of crude oil and asphalt had been discovered by the beginning of the nineteenth century. Oil produced from oil shale had been used to light the streets of Modena in Italy, paraffin wax candles had been manufactured and a small oil-shale industry had started in Scotland. The modern oil era dates from Edwin L Drake's little well in Pennsylvania, only 21.2 metres deep, which first produced oil on the 27th August 1859. A further 175 oil wells were drilled in the vicinity of Oil Creek within a year[4]. A contemporary account of those early days in the United States gave a graphic description of the congestion[5]

> ...the wells are frequently located near to the boundary of the owner's property. The object is to drain as much of his neighbour's oil as possible, for there are no partitions in the subterranean chambers corresponding to lines of surface ownership. The driller's motto is 'first-come, first-served'....

The widespread potential availability of petroleum was also recognized. The analytical chemist Tate wrote in 1863[6]

> ...petroleum is to be found in every corner of the globe. In Europe there are the English and Scottish springs; the springs of Neufchatel in Bavaria; of Amiano and St. Zelo, in Italy; Clermont and Gobian, in France; those existing in the Ionian Islands and in Sicily; and in several other parts. In Asia there are the springs of Baku, in Persia; the Rangoon springs in Burma; the Dead Sea; the springs of Hit; and in several islands of the Indian Archipelago; and in China petroleum is met with. It is also obtained from Africa.

Production at Baku in the 1870s followed a very similar pattern to the American wells and by the end of the century world production was estimated to be just under 20 million tonnes per annum, mainly shared between the United States and pre-revolutionary Russia[4]. During the 1920s the Caribbean countries Mexico and Venezuela displaced the USSR as the world's second largest producer. The Middle East developed very slowly. Although oil was first found in Iran in 1908 and Iraq in 1927, production was only 15 million tonnes by 1938, the date when the first discoveries were made in Saudi Arabia. The total world production that year was 272 million tonnes, nearly fifteen times greater than production in 1900, but this still represented less than 20% of world primary energy consumption. The dominant position of the United States, as the leading world producer and consumer had been unchallenged since 1900, with production often being up to 70% of the world total. However, the next fifteen years up to 1953, which included the second world war, saw the start of the phenominal expansion of the Middle East oil industry and the relative decline of the importance of the United States. This is shown in Table 3.2.

Table 3.2

World Oil Production and Reserves

Million tonnes

	1938				1953			
	Production		Reserves		Production		Reserves	
USA		165	2340			318	3900	
Mexico	6				10			
Venezuela	34				93			
CARIBBEAN TOTAL		34	435			103	1535	
Iran	10				1			
Iraq	4				28			
Kuwait					44			
Saudi Arabia					42			
MIDDLE EAST TOTAL		15	680			120	10570	
USSR, CHINA, EASTERN EUROPE		3	780			64	1310	
WORLD TOTAL		272	4660			658	18440	

Data from reference 7

World production more than doubled from 1938 to 1953. Although much less was known about estimating the size of an oil field at that time, the ratio of proved reserves to annual production rose from just over 16:1 to 28:1, a ratio which has changed very little over the next thirty years. In 1979 for example, when annual production was 3225 million tonnes, nearly five times the 1953 production, the ratio was still 28:1[8].

Production and Consumption

By the end of 1953 world cumulative oil production since Drake's 1859 well was estimated to be 14160 million tonnes[7]. The dramatic growth in oil production and consumption over the next twenty years up to 1973 is discussed in Chapter 2 and illustrated relative to the other main energy sources in Figure 2.5. Figure 3.1. shows how annual production has changed over the thirty year period from 1953.

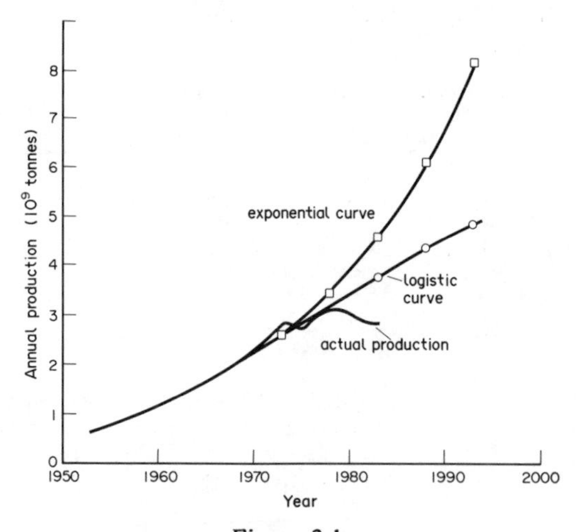

Figure 3.1

The annual production data from which Figure 3.1. was obtained[7,8,9,10] has been converted into cumulative production as shown in Table 3.3. and the two theoretical growth curves shown in Figure 3.1. were derived from this Table.

Table 3.3

World Cumulative Oil Production

Million Tonnes

Year	Cumulative Production	Year	Cumulative Production	Year	Cumulative Production
1954	14867	1964	25635	1974	48003
1955	15660	1965	27200	1975	50736
1956	16521	1966	28895	1976	53690
1957	17426	1967	30717	1977	56757
1958	18355	1968	32706	1978	59851
1959	19357	1969	34849	1979	63076
1960	20438	1970	37200	1980	66157
1961	21588	1971	39676	1981	69050
1962	22836	1972	42287	1982	71801
1963	24175	1973	45124	1983	74553

Author's derived figures from British Petroleum Company data[7,8,9,10].

The first is the exponential curve given by

$$Q_p = 14160e^{0.057t} \tag{3.1}$$

The second is the logistic equation, based on an estimated total resource Q_F of 335,000 million tonnes,

$$Q_p = \frac{335\,000}{1 + e^{3.1205-0.06316t}} \tag{3.2}$$

where t = 0 in 1953. The reasons for choosing 335,000 million tonnes are discussed later. Figure 3.2 shows how equation 3.2. was derived by plotting $\ln(Q_f - Q_p) - \ln Q_p$ against time and producing a very good straight line fit.

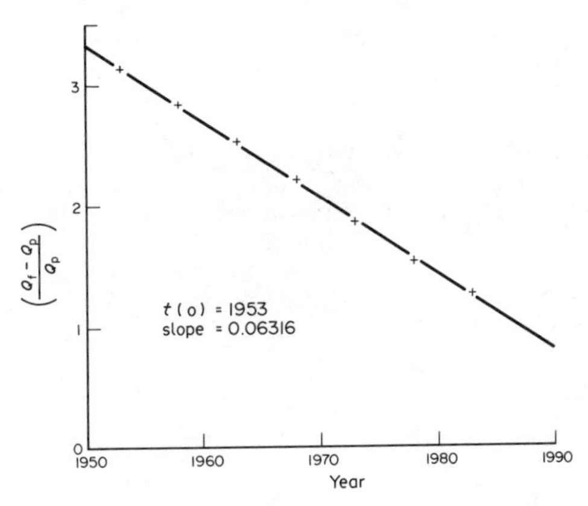

Figure 3.2

The 'super-exponential' growth from 1965 to 1973 is clearly seen in Figure 3.1. Even if the reaction of the oil-producing countries in 1973 had not taken place, it is apparent that exponential growth in the production and consumption of a resource with finite limits is quite impossible. By the end of 1983 the projected exponential growth from 1970 would have given a figure over 60% greater than actual production. The logistic curve is more realistic. Although it cannot include the effects of major disruptions caused by unpredictable price rises, the change in cumulative production from the theoretical valves is nevertheless very small, as shown in Figure 3.3. The proved reserves are taken directly from the historic data.

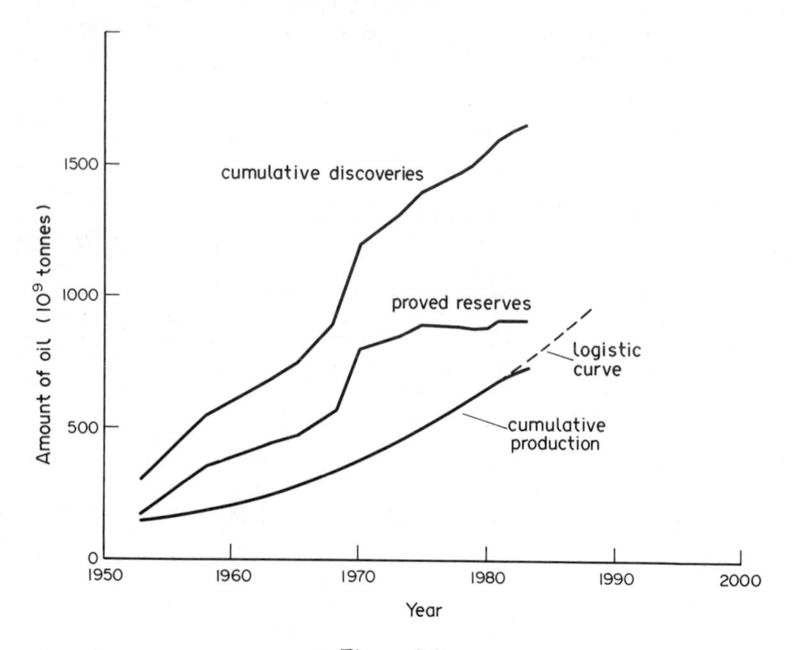

Figure 3.3

Some of the fields have been reassessed later and the earlier values have been revised upwards. Figure 3.3 could also be compared with theoretical curves discussed in Chapter 1, and shown in Figures 1.6 and 1.7. The 'cross-over' point where cumulative production exceeds proven reserves has not yet been reached, but appears to be in sight before the end of the century.

The Industrial Economies, particularly Western Europe, the United States and Japan, had already been warned of their vulnerability to incipient oil shortages prior to the 1973 price rises by Tiratsoo[4]. This can be clearly seen in Table 2.4 and 2.6 on pages 35 and 37. During the period from 1965 to 1973 they were not only consuming over 70% of the total world oil production, but their consumption was increasing at a compound growth rate of 7.4%. At the same time their relative production was declining from 45.3% of total demand to 33.9%. These trends are illustrated in Figure 3.4.

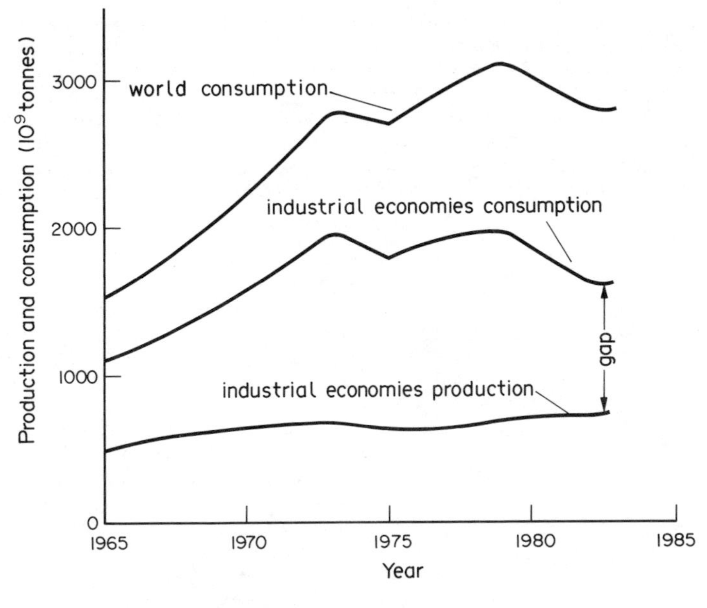

Figure 3.4

Their efforts to reassess their overall dependence on oil are discussed in Chapter 2 and resulted in the general narrowing of the 'gap' between oil consumption and indigenous oil production over the four years from 1979 to 1983 as shown in Figure 3.4. The gap would be over 100 million tonnes greater without the contribution from the North Sea.

World oil production in the major oil producing countries in selected years from 1965 to 1982, together with their proven reserves and their reserves/production ratio is given in Table 3.4. Other countries with substantial proven reserves include Abu Dhabi with 4134 million tonnes and a reserves/production ratio of 98, Libya with 2892 million tonnes and a reserves/production ratio of 52 and Nigeria with 2251 million tonnes and a reserves/production ratio of 35. The two largest producers in 1982, the USSR and the USA have the two smallest reserves/production ratios, of 14 and 8.6 respectively. Production in Mexico was little above the figures of the early 1950s until the last decade with a huge increase from the fields in the Gulf of Campeche. Venezuela's production has shown the reverse trend and has declined steadily over the decade to its levels of the early 1950s. Production in Iran and Iraq reflected the conflict since 1979, but Iran showed a partial recovery in 1983 to 129 million tonnes[11]. Both have very substantial proven reserves. Kuwait's very high reserves/production ratio was influenced by the lowest production figures for many years, but 1983 production figures were up to 45.3 million tonnes[11]. China's resources are being developed steadily. The dominating position of Saudi Arabia, with about a quarter of the world's proven reserves, is clearly shown. Table 3.4. underlines the vulnerability of the Industrial Economies to any sudden disruption to their oil supplies, as only the United States and United Kingdom have sufficient reserves to be included. The reserves/production ratio for the Industrial Economies is about 11 and their production meets about half their demand.

Table 3.4

World Oil Production 1965-82
Proven Reserves 31st December 1983
Million Tonnes

	1965	1967	1972	1973	1977	1979	1982	Proven Reserves	Reserves/Production Ratio+
USA++	387.6	437.5	470.1	457.3	409.4	424.7	430.6	3713	8.6
Mexico	17.8	20.2	24.8	26.9	53.7	80.8	149.4	6528	43.7
Venezuela	182.2	186.1	171.5	179.0	119.5	125.4	97.7	3380	34.6
Iran	95.0	129.6	251.9	293.2	283.5	158.1	98.2	6936	70.6
Iraq	64.4	60.2	72.1	99.0	115.2	170.6	48.0	5848	121.8
Kuwait	109.1	115.2	153.0	140.4	91.5	113.2	34.6	8690	251.1
Saudi Arabia	100.6	129.0	287.2	367.9	455.0	469.9	329.2	22576	68.5
USSR	242.9	288.1	400.4	429.0	545.8	586.0	612.2	8568	14.0
China	10.0	11.0	42.1	54.8	93.6	106.1	101.7	2598	25.5
UK	0.1	0.1	0.1	0.1	37.5	77.9	103.3	1788	17.3
Rest of World +	353.3	446.0	760.8	824.4	826.3	912.3	746.1	21375	28.6
WORLD TOTAL	1565	1823	2634	2872	3067	3225	2751	92000	33.4

++ Excludes Natural Gas Liquids (the liquid content of natural gas)

Data from references 8,9,10,11

+ Authors derived data

Reserves

The classification used in the previous section of 'proven' or 'proved' reserves is the only one which has been widely accepted and recognised in almost all countries. The major exception is the USSR who have a slightly different approach, with two categories at either end of the accepted 'proved' classification[12]. Proved reserves are the amounts which geological and engineering information indicates with reasonable certainity to be recoverable under the economic and operating conditions at the time when the estimate is made[10,12]. This definition and others discussed in this section apply equally to natural gas. As all other definitions such as 'probable', 'possible' and 'speculative' were capable of different interpretations, the World Petroleum Congress appointed a group of experts in 1979 to study the various classification systems and to recommend improvements. The result was two further distinctive definitions[12]. Unproved reserves refer to already discovered oil and gas deposits but their existence can only be regarded as possible or at best probable - not as reasonably certain. The expert group also suggested that unproved estimates may be given as a range. Speculative reserves come from reservoirs which have not yet been discovered. In view of the prevailing uncertainties these speculative reserves should always be given as a range.

The first step in calculating proved reserves is to estimate the total amount of oil in place in a reservoir. This is done by a series of geological studies, and geophysical tests. These include measurements of variations in the pull of gravity at the surface, variations in the earth's magnetic field and seismic surveys[13]. Large masses of dense rock increase the gravity readings and most sedimentary rocks have relatively low magnetic properties. Rock samples are taken to assess the porosity, permeability and oil content. The proportion of the oil which can be recovered from the reservoir, the recovery factor, can then be estimated. This is always an approximate calculation and has to be revised as more evidence of the nature of the oilfield is gained. For example the geological characteristics of the rocks between widely spaced wells can only be inferred, even when extensive drilling has taken place. The recovery factor also depends on several variables which can only be assessed once production has commenced. These include the rate of pressure decline and the behaviour of the oil flow through the reservoir. The faster the water rises in the reservoir as the oil is extracted the longer the pressure is likely to be maintained. Improvements in technology during the lifetime of a field could increase the recovery factor. A close watch is always kept on the price of oil. A rise in price could make it worth developing a reservoir or even a major oil field which was previously considered uneconomic.

Many estimates have been made of the world's ultimately recoverable crude oil resources from conventional sources. Some are based on the logistic curve approach favoured by Hubbert over twenty years ago[14] and discussed in Chapter 1. Others apply probability theory to estimates of how much oil there is still waiting to be discovered[15,16], while a third group pay particular attention to the political, economic and commercial factors influencing the estimates[17]. The range of estimates since 1940 is shown in Figure 3.5. The estimates up to 1970 increased almost linearly, but it is interesting to see how much the pattern is influenced by the predictions of one American analyst, L.G. Weeks.

During the mid 1970s Odell and Vallenilla[17] pointed out that the long series of estimates could be described by a highly significant regression line with a correlation co-efficient as high as 0.86. They completely rejected the generally held belief which was emerging that the total ultimately recoverable crude oil resources would be no more than 2,000 billion barrels and thought that at least double this figure was more likely. However, their regression line would have been very close to the cluster of estimates at 2,000 billion barrels in the late 1970s if Week's estimates had been neglected. During 1977 the Institut Francais du Petrole (IFP) presented a paper on oil resources to the World Energy Conference in Istambul, based on the detailed replies to

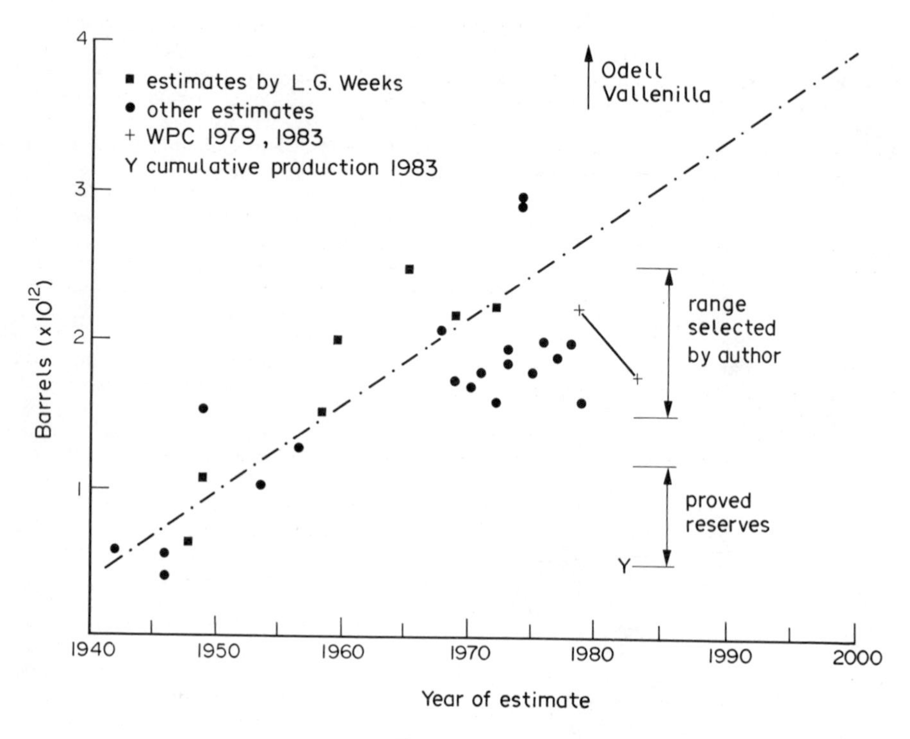

Figure 3.5

questions put to some 30 oil companies and independent analysts[18]. The average of the 22 estimates gave a total of 2,290 billion barrels of oil, or about 305 billion tonnes (using the conversion specified by IFP of 7.5 barrels per tonne). This figure was confirmed by Halbouty and Moody[15] at the World Petroleum Congress in 1979, where they gave a 90% probability that the undiscovered potential was at least 40 billion tonnes and a 10% probability that they could be as high as 345 billion tonnes, from which they postulated the existence of 141 billion tonnes. These assumptions were based on the arbitrary assumptions that the recovery factor, then between 25 and 30%, could be increased to 40% and that there could be offshore discoveries down to a depth of 2000 metres. The British Petroleum Company[13] considered both assumptions to be over-optimistic, particularly as many recovery factors in existing reserves estimations were already in the range from 35 to 40%, the Middle East had already provided about three-fifths of all world discoveries in the period from 1945 to 1977 and that about three quarters of all world oil reserves ever discovered lie in some 280 giant fields. There are relatively few geological regions in the world that can contain a giant field and most have been adequately explored. Confirmation of the British Petroleum view come in the 1983 World Petroleum Conference. The paper presented by Masters, Root and Dietzman[16] reassessed the undiscovered potential with a 95% probability that 46 billion tonnes to be discovered but only a 5% probability that they could be as high as 202 billion tonnes, producing an ultimately recoverable total of 245 billion tonnes, of which only 78 billion tonnes were undiscovered potential, and the sharp fall in the estimate over the four year period shown by the solid line in Figure 3.5. To achieve an undiscovered potential which would lift the ultimately recoverable total to over 3000 billion barrels would need the discovery of another Middle East and their studies indicated no possibility for such an occurence[16]. Two other factors which are considered in the analysis of trends in discoveries are the amount of oil found per exploratory metre drilled and the size of the oil fields discovered[19]. Both of these indicators have also been declining, confirming Tiratsoo's comment[4] that an enormous drilling programme would be necessary to bring the general intensity of exploration in other parts of the world up to anything like the present(1972) American pattern. Even if this were to be physically and economically possible he felt there must be considerable doubt as to whether it could result in a proportionate increase in the new oil reserves which would be discovered.

Oil and the United Kingdom

The history of North Sea oil dates from the discovery of the Forties field in 1970 and the early period up to 1980, when the UK first produced more oil than was consumed, is discussed in Chapter 2. The discoveries were quite unexpected. Only six years before drilling commenced for oil and gas in 1964 one of the worlds leading oil economists, Dr. P.H. Frankel, commented[20]

> The question whether oil is and will be available in sufficient quantities is a very pertinent one, particularly critical for us because unfortunately the United Kingdom is apparently devoid of sizable indigenous oil reserves and thus, having to rely on imports, our supply problems are not domestic but international...

The British National Oil Company (BNOC) was created by the Government in 1976 to advise the Government on all matters relating to oil and to act as an oil company. It inherited shares in certain North Sea blocks previously owned by the National Coal Board and was given powers to re-negotiate licences which gave them access to just over half the North Sea production. Nevertheless there have been strong criticisms of Government policies on taxation, which meant that virtually no new development work in the North Sea took place over the two years up to February 1983[21] although oil exploration continued at an increased level. The new development work is essential because of the long time needed to bring an oil field into full operation.

Since 1980 annual production has continued to rise, reaching 112 million tonnes by the end of 1983[11] and is expected to reach a peak of about 125 million tonnes in 1985. Thereafter production from the first generation of fields and from other fields already being developed was expected to decline and could range between 60 to 95 million tonnes per annum by 1990[21]. The Government eventually accepted that tax remissions were needed if the new fields were to be brought into production in the early 1990s and these were announced in April 1983[22]. Estimates by the Department of Energy of discoveries still to be made in the North Sea suggest that these might lie between 800 and 1800 million tonnes. These new fields will almost certainly be small and more expensive to develop.
North Sea oil has provided the UK with secure oil supplies, possibly to the end of the century and has, in the words of Sir David Steel, former Chairman of The British Petroleum Company[23], given the country

> a great opportunity to retrain its people and re-equip its industry, so that it can be competitive when the time comes for imports of energy (not necessarily of oil) to be resumed on a considerable scale...we need to learn to use energy more efficiently, as our competitors are doing, and to develop the knowledge and techniques necessary to move steadily to new energy sources as the old are depleted...this opportunity is not likely to be offered to us twice.

The Multinational Oil Companies and OPEC

The Organization of Petroleum Exporting Countries (OPEC) was formed in September 1960 as a direct result of the actions of the major multinational oil companies in reducing the prices upon which they had agreed to pay royalties and taxes to the exporting countries. The OPEC group consists of thirteen countries, Saudi Arabia, Kuwait, United Arab Emirates, Qatar, Iran and Iraq in the Middle East, Algeria and Libya in North Africa and Ecuador, Gabon, Indonesia, Nigeria and Venezuela. Between them they held some 65% of the world's proven reserves at the end of 1983[11]. The most influential country in the OPEC group is Saudi Arabia, with enormous reserves and production capability combined with great financial wealth and a relatively small population. This makes Saudi Arabia unique in that it has long term interests in maintaining political influence through its oil markets, retaining its position as the world's leading oil exporter, and conserving its oil reserves while protecting its own position within the world economy[23]. Since the mid 1970s Saudi Arabia has tried to promote the policy that oil is a global issue and should be dealt with on that basis. The seven major multinational oil companies are collectively known as the 'Seven Sisters' and consist of Exxon, Royal Dutch/Shell, British Petroleum, Texaco, Mobil, Socal and Gulf Oil. They effectively controlled world oil production from the 1920s until after October 1973.

Following the closing of the Suez Canal in the Six-Day war the first of the oil shortage crises occured. Several countries within OPEC were able to obtain a substantial rise in their tax payments on their oil exports. The sequence of events from October 1973 has been covered in Chapter 2. This resulted in a series of price rises in the posted price of Arabian Light crude oil from 2.6 US dollars per barrel during the early part of 1973 to a peak of 35 US dollars in 1981, followed by a fall to just below 30 dollars by the end of 1983 (prices not corrected for inflation). Relative to the price paid by the consumer for a typical OPEC barrel in Western Europe, the OPEC producers were receiving 19% of final price in 1973 and 54% in 1982.

Shale Oil

Oil shale is a fairly common thinly layered sedimentary rock containing solid organic material, the hydrocarbon wax kerogen. At least twenty other additional names for oil shales have been found in the literature[24] including 'the rock that burns'. As recently as the early 1950s shale oil was being produced at a level of 100 000 tonnes a year in Scotland, but the operation closed in 1962[1]. Two countries, the USSR and the People's Republic of China, were successfully operating major oil shale developments in the early 1980s[24]. In the United States, the Green River Formation in parts of Colerado, Utah and Wyoming is the largest and best known oil shale with, as yet, very limited commercial development[23]. Oil shale does not contain oil, but if it is broken up and heated in a retort to temperatures greater than 480°C, the kerogen decomposes, giving off an oily vapour which condenses to a viscous shale oil containing up to 80% of the original energy in the kerogen. Some 6% of the original energy in the kerogen is also given off as gas, the remainder being a carbonaceous residue. Because this shale oil contains considerable amounts of nitrogen compounds and sulphur it needs further processing before a 'syncrude' product similar to crude oil is obtained[23]. There is no minimum content of organic matter or oil yield which distinguishes oil shale from other sedimentary rocks.

Oil shale deposits are found throughout the world. However, there are many obstacles to the large scale exploitation of oil shales, the main one being the comparatively small amount of oil associated with the shale. This was put very succinctly by Foley[1].

> Oil shales cannot usually be mined by open-cast methods....shale for retorting has to be quarried out of deep workings or mined in much the same way as coal. Only the richest deposits are worth working in this way. A tonne of coal is a tonne of coal. But a tonne of oil shale is a lot of rock and a little oil which has to be retorted and upgraded. Obtaining the same amount of energy from an oil shale which yields a barrel or oil per tonne (159 litres) requires five times as much mining as coal.

If only 15% of the world's oil consumption in 1983 were to be replaced by rich shale yielding a barrel of oil per tonne,, the amount which would need to be mined would be equal to the output of the entire world coal mining industry.

The alternative to mining is to process the shale fully or partially in situ. This would avoid the problems of handling and disposing of very large volumes of rock, but the basic problem is the impermeability of the oil shale beds. Work in the United States has included drilling a pattern of boreholes followed by fracturing the shale hydraulically or by explosives. Heat is then applied, either by establishing combustion in the shale or by injecting hot liquids or gases. In a second approach, some shale is mined, and a broadly similar fracture and heating process is then followed. Although these methods can reduce or even eliminate the problem of the disposal of spent shale, a new set of process heating problems in the fractured shale have still to be solved[23].

There are no 'proven reserves' for either oil shales or tar sands as so much depends on the development of suitably economic extraction technologies. The most widely quoted estimates for the resources of oil shale come from Duncan and Swanson[25]. Neglecting any resources estimated to lie in range from 5 to 25 gallons/ton, (the gallon/ton unit is equivalent to 4.17 litres/tonne), the richest part of the known and evaluated world shale oil resources in the range from 25 to 100 gallons/ton are shown in Table 3.5.

Table 3.5

World Shale Oil Resources

	Billion Barrels	
	Known and evaluated	**Speculative**
Africa	100	4000
Asia	90	5500
Australia/New Zealand	Small	1000
Europe	70	1400
North America	600	3000
South America	50	2000
Total	910	17000

Data from reference 25.

The world total known and evaluated resources of 910 billion barrels or 124 billion tonnes of oil are about one third larger than the proved resources of crude oil at the end of 1983. In the United States the Mahogany Ledge region of the Green River Formation has known and evaluated reserves of about 14 billion tonnes of oil, just under three times their proved conventional crude oil reserves. Even if only half the speculative resources in Table 3.5 are eventually proved, total world oil reserves would increase to over three hundred years of consumption at the level experienced in 1983.

Tar Sands

The origin of the tar sands is not clear. One theory is that the sands were once on the ancient sea shores, were scattered by the winds and became impregnated with very viscous or even solid petroleum. This petroleum material is usually called bitumen. The tar sands do not contain tar, which a manufactured product. While tar sand deposits are distributed throughout the world, there are only a few rich areas which have been identified and partially exploited. The two largest are the Athabasca Tar Sands in Alberta, Canada and the Orinoco Heavy Oil Belt in Venezuela, with smaller resources reported in the USA and the USSR[24]. The Athabasca development is by far the most advanced in the world, with commercial production commencing in 1967, although a demonstration road surface using the bituminous material was laid as early as 1915[1]. The sands are mined by conventional surface strip-mining methods followed by a series of separating and cleaning processes. Production at Athabasca had approached 200 000 barrels per day by the early 1980s with forecasts of a fourfold output by 1995[23]. This would be equivalent to nearly 40 million tonnes per annum. The environmental problems associated with tar sands are formidable as about two tonnes of sands are needed to produce a barrel of oil. An output of 40 million tonnes of oil would create nearly 600 million tonnes of waste material.

A recent estimate of the world resources of oil from the tar sands[26] gave a range between 340 billion and 820 billion tonnes, but some 90% of these resources cannot yet be used as they are below the maximum depth for surface mining. In situ production is the only possibility and two general methods have been proposed, combustion and hot water or steam injection. In the combustion processes, a fire front is established and travels through the tar sand formation towards the production well(forward combustion) or the air injection well(reverse combustion). Controlling either process is very difficult. Steam injection lowers the viscosity of the petroleum to a point where it can flow easily to the surface. Both techniques are similar to tertiary recovery methods for crude oil and the in situ proposals for the shales. By the end of 1983 estimates for the cost of producing a barrel of oil from the tar sands and oil shales of North America were in range from 40 to 85 United States dollars, compared with the production cost of about 5 dollars for Middle East oil[27].

Some Environmental Issues

All the operations of the oil industry have some effect on the environment. These can range from possible damage to wild life in exploration or production areas, the rehabilitation of land after pipe laying or mining operations, atmospheric emissions from refineries and chemical plants to the comparatively rare sea incidents such as the 'Torrey Canyon', 'Amoco Cadiz' or Bantry Bay disasters experienced in Great Britain and Ireland in recent years. At present about one fifth of the oil produced in the world comes from offshore fields. Estimates indicate that about one third of future crude oil production could come from the Continental Shelf at depths of less than 200 metres and one third from deep water and polar regions[27]. Adverse environmental problems associated with offshore fields include the possible contamination of sea beds and fishing shoals, damaging the fishing industry and the pollution of vast areas of beaches, problems also associated with leaks or spills from oil tankers. Experience gained during the 1970s and early 1980s has suggested that the effects of oil on the marine environment are not irreversible and new techniques for recovering and dispersing oil spills have been developed. A wide battery of protective devices are associated with pipelines on land to detect leakage. Refineries use very considerable quantities of cooling water which is now purified and returned to rivers and estuaries in much the same or even better condition. A leading company[28] warns that whilst the objective of improving the environment may be highly desirable, measures to achieve the required objectives could involve high expenditure without any guarantee of success.

Summary

The period of exponential growth in oil production and consumption experienced since the early 1950s which resulted in its dominating position in the overall energy market has ceased. The oil industry has started the reassessment which could lead the use of oil back to those special applications which characterized its growth up to the 1940s - its use as a petrochemical feedstock, as a lubricant, its special uses in many forms of transport and as a flexible product in the higher value areas of manufacturing. Although there are other known routes to oil such as the oil shales and tar sands, these are expensive and no simple replacement has yet appeared.

The crucial difference between the leaders of the oil industry 100 years ago and today is that the early pioneers anticipated that they were seeing the beginning of the oil era. Today they are seeing the beginning of the end of the oil era[29]. In a curious way their perceptions of the future are very similar. The early pioneers could only have a very tentative and barely understood vision of how the industry could develop. Today the end of the oil era is equally uncertain.

References

1. Foley, Gerald, The Energy Question, Second Edition, Penguin Books Ltd., 1981.

2. Facts about Oil, The American Petroleum Institute, 2101 L Street, N.W., Washington, D.C., 20037.

3. The Petroleum Handbook, Fourth Edition, Shell International Petroleum Company, London, 1959.

4. Tiratsoo, E.N., Oilfields of the world, Scientific Press Ltd., Beaconsfield, Bucks, 1973.

5. Harpers New Monthly Magazine, 1885, quoted in '90 Years On', Esso Magazine, 108, 7, 1978.

6. Tate, A.N. Petroleum and its products, 1863, quoted in '90 Years On', Esso Magazine, 108, 6, 1978.

7. B.P. Statistical Review of the World Oil Industry, The British Petroleum Company Ltd., London, 1963.

8. Ibid, 1980 edition.

9. Ibid, 1973 edition.

10. B.P. Statistical Review of World Energy 1982, The British Petroleum Company, London, 1983.

11. Worldwide Report, Oil and Gas Journal, 77-109, December 26, 1983.

12. Martinez, A.R., Ion, D.C. et al., Classification and Nomenclature Systems for Petroleum and Petroleum Reserves, World Petroleum Congress, London, 1983.

13. Estimating world oil reserves, Briefing Paper, The British Petroleum Company, June 1980.

14. Hubbert, M. King, Energy Resources: A Report to the Committee on Natural Resources, National Academy of Sciences - National Research Council, Publ. 1000-D, Washington D.C., 1962.

15. Halbouty, M.T. and Moody, J.D., World Ultimate Reserves of Crude Oil, World Petroleum Congress, Bucharest, 1979.

16. Masters, C.D., Root, D.H. and Dietzman, W.D., Distribution and Quantitative Assessment of World Crude Oil Reserves and Resources, World Petroleum Congress, London, 1983.

17. Odell, Peter R, and Vallenilla, Luis, The Pressures of Oil, Harper and Row, London, 1978.

18. World Energy Resources 1985-2020, 1-48, World Energy Conference, 1977.

19. Steinhart, John S., and McKellar, Barbara J., Future availability of oil for the United States in Perspectives on Energy, Third Edition, Oxford University Press, 1982.

20. Frankel, P.H., Oil resources and economics, Proc. Conf. Economic Aspects of Fuel and Power in British Industry, (5-7 November 1958), 9-18, Manchester University Press, 1960.

21. UK North Sea taxation, Briefing Paper, The British Petroleum Company, February 1983.

22. Lawson, Nigel, Stability restored to oil market, Department of Energy, Press Notice, April 14, 1983.

23. Harder, Edwin L., Fundamentals of Energy Production, John Wiley and Sons, New York, 1982.

24. Smith, J.W., Synfuels: Oil Shale and Tar Sands, in Perspectives on Energy, Third Edition, Oxford University Press, 1982.

25. Duncan, D.C. and Swanson, V.E., Organic Rich Shale of the United States and World Land Areas. U.S. Geological Circular 523, Washington D.C., 1965.

26. Gray, G.R., Oil Mining, Oil Sand, in Encyclopedia of Energy, McGraw-Hill, New York, 1981.

27. Shell Briefing Service, Number Five, Shell International Petroleum Company, London, 1983.

28. Protecting the Environment, Briefing Paper, The British Petroleum Company, April, 1983.

29. Hamilton, Adrian, Purveyors of Energy, Esso Magazine, 108, 2-5, 1978.

Further Reading

A useful well-illustrated introduction to the science and technology of petroleum is given by Dr Alfred Mayer-Gurr's Petroleum Engineering, Pitman Publishing Ltd., 1976,

Volume 3 of Geology of Petroleum, while Dr. E.N. Tiratsoo's Oilfields of the World, Scientific Press, 1973, covers the whole field in much greater depth. A very detailed and authoritative account of the North Sea developments, including advanced recovery methods and estimates of the reserves in each field, is in North Sea Oil. by Kenneth Klitz, Pergamon Press, 1980.

The classical book on oil economics is Dr. P.H. Frankel's Essentials of Petroleum, Frank Cass, London, first published 1946 with a new edition in 1969.

Economic and political issues are covered in Peter Odell's Oil and World Power, first published by Penguin Books in 1970, and in many editions since then. Odell joins with Luis Vallenilla in The Pressures of Oil, Harper and Row, 1978, which explores the possible collaboration of the OPEC and OECD countries. Oil Politics in the 1980s, by O. Noreng, McGraw-Hill Book Company 1978 covers international co-operation.

In the UK, the Briefing Papers and Annual Statistical Review of World Energy, from the British Petroleum Company, and Shell Briefing Service and the Esso Magazine, are very useful information sources. Among the journals the Oil and Gas Journal and the Petroleum Economist are particularly good, with both statistical data and comment on current events. The latest technical papers come from the proceedings of the 1983 World Petroleum Congress, London.

Exercises

1. Two estimates for the ultimately recoverable crude oil resources Q_f are 272 billion tonnes (2000 billion barrels) and 204 billion tonnes (1500 billion barrels). By plotting the graph of $\ln (Q_f-Q_p)/Q_p$ against time, as in Figure 3.2., show that the higher estimate could give the logistic curve

$$Q = \frac{272\ 000}{1 + e^{3.1205-0.0666t}}$$

   At what point would half this resource base be depleted? Show that the lower estimate does not appear to satisfy a straightforward logistic curve.

2. The following table shows the ratio of proven world oil reserves to annual production every five years from 1953 to 1983.

Year	Ratio of proven reserves to production
1953+	28.0
1958+	40.2
1963+	33.7
1968	45.9
1973	30.9
1978	28.6
1983	32.6

   \+ Figures based on known size of field at that time.

   (a) Plot a curve of 'final depletion date' against the actual date on which the ratio was calculated.

   (b) What are the limitations to any conclusions which could be drawn from this curve?

3. World oil consumption in 1983 was 2821 million tonnes, proven reserves were 92 thousand million tonnes and cumulative production was 74 266 million tonnes. As part of a medium term examination of future energy trends the following assumptions were made:

   (1) World oil consumption (and production) will increase by 50 million tonnes per annum each year for the next ten years.

   (2) New reserves are proved at a rate of 2000 million tonnes per annum for five years. In the sixth year a large field with a potential resource of 18 000 million tonnes is discovered. For the rest of the ten year period, new reserves are proved at a rate of 1600 million tonnes per annum.

   (a) Show the projected relationship between cumulative discoveries, proved reserves and cumulative production over the ten year period to 1993 on a diagram similar to Figure 3.3. Is this a probable projection, or a possible projection or very unlikely?

   (b) What would the ratio of proven reserves to production be in 1993 with these assumptions?

4. During the late 1970s the Saudi Arabian Oil Minister gave repeated warnings that the Western countries must change their life styles and save energy. In 1979 two comments on the shortages of petrol experienced then in the United States were as follows:

   'Travelling is no longer a luxury. It's a need, a right. You've got to get out of the house, get away from the urban centres, and people are going to get away one way or another. Many Americans think of their car as a second home - a castle'. Professor Edward Mayo, Notre Dame University.

   'The car is America's magic carpet and it give people freedom and autonomy - it's their little box where they have control over their environment. There is tremendous resistance to anything that threatens the use of the car'. Professor Wayne Youngquist, Marquette University.

   To what extent have the Western countries changed their life styles and started to save energy since the end of the 1970s? Does the car still have the same significant position in American society today that it appeared to have in 1979 when these comments were made? Would these comments be equally true in the UK during the 1980s?

5. In January 1984 a report on the Iran-Iraq war drew attention to the problems which could result from the closure of the Straits of Hormuz through which most of the oil from the Middle East fields is transported in super-tankers.

   (i) How much oil passed through the Straits in 1983?

   (ii) How many days of oil reserves have the countries of the Industrial Economies in stock?

   (iii) What effect would the closure of the Straits of Hormuz have on the world economy if the closure lasted for (a) three months (b) three years. In each case assume production can resume at a level of 600 million tonnes per annum.

6. Examine the Annual Reports of any major oil company. Have they diversified their interests into other energy activities since the 1970s and, if so, what has been the result of this diversification?
   Which of the renewable energy resources has attracted most interest from the major oil companies?

7. The United Kingdom is experiencing a temporary period of energy self-sufficiency, with exports of North Sea oil exceeding the volume of oil imports. What are the latest forecasts for North Sea oil resources? When could the period of net oil exports end? What impact could this have on the UK economy?

8. If a Government decision to reduce oil consumption in your country by a total of 25% over the next two years was announced today, what suggestions could you make to the Department of Energy to help the country reach this target?

9. In the introduction to his book 'Oil and World Power', Penguin Books Ltd., Fifth Edition, 1979, Peter R Odell stated ".... a day rarely passes without oil being in the news." What news items relating directly to oil have appeared in the past two months?
   Do these modify any of the broad conclusions reached in the Chapter?

10. Is there any evidence that oil from shale oil can be produced on a commercial basis? Discuss the main environmental problems associated with both oil-shale exploitation and the production of oil from tar sands.

Answers

1. About 1998.
2. (b) There are many factors influencing the reserves/production ratio in this period, for example the 1973 and 1979 oil price rises and the subsequent world recession. It is very difficult to draw any firm conclusions from this curve.
3. (a) Possible (b) 28.8

4. COAL

Introduction

Coal is a hard, opaque, black or dark coloured combustible sedimentary rock. Coal deposits were formed from the remains of plant materials by the combined action of heat and pressure. The process of change from dead vegetation to coals is known as coalification. The coalification process started with some of the dead vegetation being preserved from rapid decay either by being submerged in swamps or estuaries or under successive layers of vegetation. This slow decaying process occurred largely without air and led to the formation of peat. Coalification then followed through the combined action of heat and pressure, first to brown coal, then lignite, subbituminous coal, bituminous coal and anthracite. Brown coal, lignite and subbituminous coal are known as the 'soft' coals, while bituminous coal and anthrancite are known as the 'hard' coals. Most bituminous coals date from the Carboniferous period, some 280 to 345 million years ago, while the lignites and brown coals are relatively younger, dating from the Tertiary and Cretaceous periods, from 3 to 136 million years ago[1]. Today's peat deposits were formed during the past million years.

Coal is chemically and physically very complex. Pure coal consists mainly of carbon, with lesser quantities of hydrogen, oxygen, nitrogen and sulphur. During coalification the relative proportions of these elements change. The carbon content increases as some of the other elements are reduced. Where coals are decomposed by the action of heat, or pyrolysis, the hydrogen and oxygen content is practically eliminated as the volatile matter is driven off. Typical volatiles are carbon monoxide, carbon dioxide, methane and water. Fixed carbon is the combustible residue left behind when the volatile matter has been driven off, and normally represents that portion of the fuel that must be burnt in the solid state. There are a number of different ranking systems for coal. These include the classifications of the National Coal Board in the UK, the American Society for Testing Materials and the International Standards Organisation. Some typical values, in descending order of rank, are given in Table 4.1. Descending rank is associated with a reduction in fixed carbon and decreasing calorific value.

Table 4.1

A Classification of Coal by Rank+

Class	Carbon %	Hydrogen %	Fixed Carbon%	Calorific value MJ kg^{-1}
Anthracite	95-98	2.9-3.8	91-95	>32.5
Low volatile bituminous	91-92	4.2-4.6	80-85	>32.5
Medium volatile bituminous	87-92	4.6-5.2	70-80	>32.5
High volatile bituminous	82.5-87	5.0-5.6	60-70	26.7-32.5
Subbituminous	78-82.5	5.2-5.6	55-60	19.3-26.7
Lignites	73-78	5.2-5.6	50-55	<19.3

+ On a dry mineral-matter free basis. Data from Williams, 1981[2].

For a simpler classification, the World Energy Conference[3] has adopted the categories of hard coal, defined as any coal with a heating value above 23.76 MJ kg^{-1} on a moisture and ashfree basis and brown coal for any coal, including subbituminous coal, with a lower heating value than 23.76 MJ kg^{-1}. But even this simple classification is not followed in all countries. For example in Canada and the United States 'brown coal' normally refers only to lignite and excludes subbituminous coal.

Coal Resources, Reserves, Production and Growth

The early history of the modern coal industry was dominated by two countries, the United States and the UK. By 1885 coal had overtaken wood as the primary energy source in the United States and by 1900 they had taken the lead as the world's main coal producer. The growth of world coal production and the relative decline of the UK as a major world producer is shown in Table 4.2.

Table 4.2.

Coal Production 1880 - 1980

Year	World Production (million tonnes)	UK Production (million tonnes)	UK as % of World*
1880	320	149.4	46.7
1900	780	228.8	29.3
1920	1200	233.2	19.4
1940	1700	227.9	13.4
1960	2100	196.7	9.4
1970	2450	147.1	6.0
1975	2565	128.7	5.0
1980	3030	130.1	4.3

Data from references 4, 5 and 6

* Derived figures

The initial growth rate in world production averaged about 4.5% per annum up to 1900, but from the end of World War I to 1960 growth rates declined fairly steadily, with the period from 1940 to 1960 averaging only 1% per annum. There were several reasons for this. The world depression in the 1930s reduced demand. Improved efficiences in electricity generation and in other major coal using industries such as steel also curbed growth. Finally there was the major penetration into a number of traditional coal markets, such as domestic and industrial heating, of oil and natural gas. In many countries however, coal retained its hold on electricity generation and it was not until the middle of the 1960s that oil became the world's main prime energy source.

Coal resources are at least five times greater than the combined resources of all known or possible oil and gas resources. Many different resource classification systems have been developed. These vary from one country to another and even from one coalfield to another in the same country. The most commonly used classification system was defined by the World Energy Conference in 1978[3]. Geological Resources of coal were classified as a measure of the amount of coal in place. Technically and Economically Recoverable Reserves were classified as a measure of the quantities that can be economically mined with current mining technology and at current energy prices. Several detailed assessments of world coal reserves and resources were carried out between 1977 and 1980[3,7,8,9]. Coal Worldwide[8] was published in five volumes in 1980 and contains statistics from 2,370 single coal deposits in 69 countries. Table 4.3. gives the resources, reserves and production for the major coal-producing countries. The projected annual growth figures were derived from data given in the major analysis of the world's energy and coal prospects to the year 2000, the World Coal Study (WOCOL), published in 1980[9,10]. The reserves/production ratio for the world shows a figure of nearly 350 years, based on 1977 production.

The ten countries shown in Table 4.3. account for 98% of the geological resources and 92% of the recoverable reserves. The USSR, the United States and the People's Republic of China have about 60% of the recoverable reserves when the substantial brown coal deposits in the USSR are added.

The data in Table 4.3 was reassessed in 1980 by the World Energy Conference and classified by rank as shown in Table 4.4. More precise definitions of Proved Reserves, Additional Resources and Proved Recoverable Reserves were given:

> Proved Reserves represent the fraction of total resources that has not only been carefully measured but has also been assessed as being exploitable in a particular nation or region under present and expected local economic conditions (or at specified costs) with existing available technology.
>
> Additional Resources embrace all resources, in addition to proved reserves, that are of at least foreseeable economic interest. The estimates provided for additional resources reflect, if not certainty about the existence of the entire quantities reported, at least a reasonable level of confidence. Resources whose existence is entirely speculative are not included.
>
> Proved Recoverable Reserves are that fraction of proved reserves in place that can be recovered, or extracted from the earth in raw form, under the above economic and technological limits. The ratio between these two is commonly called the recovery factor.

Table 4.3

World Coal Resources, Reserves, Production and Cumulative Production (1977-2000) for major coal-producing countries (m.t.c.e.)

	Geological Resources	Technically and Economically Recoverable Reserves	Production		Estimated Cumulative Production 1977-2000	Cumulative Production as a Percent of Reserves (%)
			Actual 1977	Projected 2000		
Australia	600,000	32,800	76	326	4,200	13
Canada	323,036	4,242	23	159	1,800	+
People's Republic of China	1,438,045	98,883	373	1,450	20,000	20
Federal Republic of Germany	246,800	34,419	120	150	3,100	9
India	81,019	12,427	72	285	3,900	31
Poland	139,750	59,600	167	313	6,700	11
Republic of South Africa	72,000	43,000	73	228	3,300	8
United Kingdom	190,000	45,000	108	162	3,000	7
United States	2,570,398	166,950	560	1,883	25,000	15
Soviet Union	4,860,000	109,900	510	1,100	18,000	16
Other Countries	229,164	55,711	368	724	14,000	25
Total World	10,750,212	662,932	2,450	6,780	103,000	16

+ The published estimates of Canadian reserves are not comparable in the WOCOL study

Data from the World Coal Study (WOCOL), references 9 and 10

Table 4.4

World coal resources by rank (1980)
10^9 tonnes of coal equivalent

	Anthracite and Bituminous	Subbituminous	Lignite	Total
Proved Reserves in place	775	173	113	1061
Additional Resources	6161	2991	848	10000
Total	6936	3164	961	11061
Proved Recoverable Reserves	488	112	88	688

Data from reference 11

The total of proved recoverable reserves is within a few percent of the estimate in Table 4.3.

Recovery factors are not widely published. Factors ranging from 95% to 12% for bituminous coal and anthracite have been given[12]. Although various technical reasons dictate why 100% recovery factors are almost impossible, market demand at the time of extraction and the availability of other competing energy sources are often more important influences. If only part of the proved recoverable reserves are required at one particular time, mining methods could be chosen to allow the remaining unworked coal to be extracted in the future.

As the coal resource base is so great it is often difficult to obtain direct information about proved reserves and possible additional resources. Table 4.3. shows that the major coal producing countries have between 150 and 500 years of proved reserves, based on 1977 production. These figures are unlikely to change significantly over the next decade as growth in production was well under 3% per annum up to 1982. There are also problems in interpreting the information, as there is no internationally accepted ranking for coal. These problems were experienced during the preparation of Table 4.3., where some national data differed from World Bank data and different sets of conversion figures were used for the "tonne of coal equivalent". The compiler of the resources data for the 1980 World Energy Conference, Dr. J. Koch[11], commented

> Only 42 of the countries possessing coal or peat deposits returned the questionnaires. This is about half of the countries currently mining or who have previously mined these fuels. Some of the questionnaires were incompletely answered....
>
> ...Even though the structural characteristics of coal deposits are relatively well understood, there are still problems in preparing statistics for the resources. One source of uncertainty and differences between statistics prepared by different sources is the fact that the subdivision of coal according to rank is not internationally unified. A more serious problem is the lack of unification in the sub-divisions and nomenclature of resource categories drawn up in relation to confidence level of geological predictions, thickness and depth of seams, and technical and economic recoverability....

Coal Mining Methods

The two main methods of mining coal are surface or open casting mining and underground or deep mining. The choice of method is mainly determined by the geological characteristics of the coal deposit. Only deposits occuring at shallow depths are likely to be surface mined. Occasionally, when the coal occurs in multiple seams, it becomes possible to use surface mining techniques at considerable depths.

Surface Mining

Surface mining has a number of advantages over underground mining. The economics are often very favourable and a higher proportion of the coal in place can be recovered. Productivity is higher as surface mining is less labour-intensive. The labour force can often be recruited directly from those with experience in similar industries, such as construction or transport. Economies of scale in production have been achieved in recent years with the trend towards increased unit size in equipment such as draglines and high capacity conveyors. There are a number of environmental objections to surface mining, which are discussed later, but in spite of these, surface mining is expected to supply the major proportion of the coal produced in the two largest producing countries, the USSR and the United States, and in many other countries including Australia and Canada. One projection for the United States[10] suggested that surface mined output could more than double to 850 million tonnes per annum by the year 2000.

Underground Mining

Underground Mining is normally carried out by either the longwall method or the room-and-pillar method. In the longwall method, coal is extracted across the whole cutting face of a seam, using shearer-loaders in conjunction with fully mechanized face support systems. Two parallel main passages, some 200 metres apart, are steadily driven out. At the same time the working face, which is at right angles to the main passages, is also being advanced. Hydraulically powered supports are used to hold up the roof in an access passage behind the working face while the coal is cut and removed by a conveyor belt system. The supports are then advanced, allowing the roof to fall behind them into the section which formed the access passage for the previous cut. The new access passage is once more fully supported and in the position from which coal has just been removed. By allowing the whole roof to fall, a relatively uniform subsidence occurs. This minimises the impact at the surface. It is not widely appreciated that the face workers spend their shifts in the access passage under the roof supports. If the coal seam is very shallow the headroom may only be in the order of one metre. This is illustrated in Figure 4.1., which shows conditions at the coal face of a particularly shallow seam. In these conditions the only way to get to the cutting machines is to crawl - often over a rough mixture of coal, debris, sharp stones and water[13].

The longwall method is used in most Western European countries to withstand the high rock pressures caused at great depths. It allows a higher proportion of the coal to be extracted than the room-and-pillar method. Productivity is often higher because more sophisticated automatic machinery can be used.

In the room-and-pillar method, the coal is extracted at relatively shallow depths by leaving pillars of coal in place to support the roof. This considerably lessens the possibility of subsidence at the surface but the method leaves some 40% to 50% of the coal in the mine. Any subsequent attempt to recover more coal by removing the pillars could cause severe subsidence. This method is more common in the United States, but is gradually being superceded there by the longwall method as working depths increase.

Figure 4.1

The Transport of Coal and WOCOL

One of the most important factors influencing the use of coal is the transport system. This was recognized in the major international analysis carried out by experts from 16 major coal-using and producing countries, published in 1980 as the World Coal Study (WOCOL)[9,10]. In this study, the production and use of coal was projected to grow by a factor of at least 2.5 over the twenty year period to the year 2000. World trade in steam coal was projected to grow 10 to 15 fold within the same period. This would require major capital investments in new and expanded coal mining facilities, in coal-fired power stations and in coal conversion systems. New transport systems for the export and import of coal as well as internal transport in both coal producing and coal using countries would be needed. The links in the system from mines to users were visualized within the context of a coal chain, as shown in Figure 4.2, to focus attention on the possible delays which could occur if transport developments are not fully integrated in the overall planning.

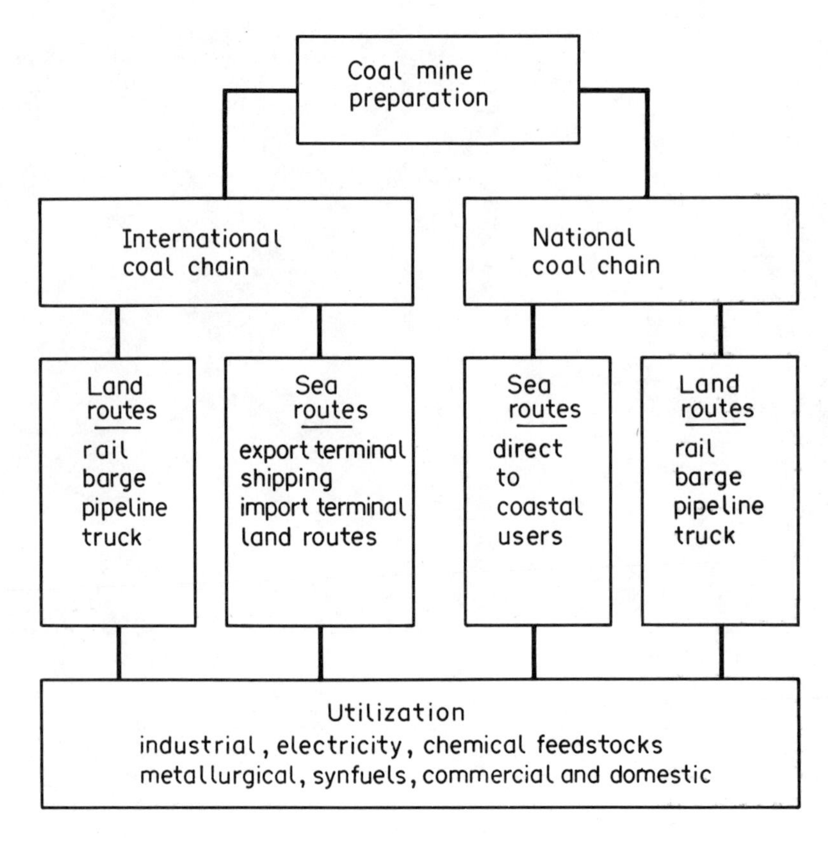

The Coal Chain

Figure 4.2

Coal in the UK

The early history of the coal industry in the UK was described in Chapter 2. By the beginning of the twentieth century the great size of the resource base had been appreciated. Figures broadly similar to today's estimates of Geological Resources of 190 000 million tonnes (see Table 4.4) had already been forecast. In 1905 the Report of the Royal Commission on Coal Supplies described the contents of the proved coal fields of the United Kingdom as 100 000 million tons with a further 40 000 tones in still unproved fields. Taking a round figure of 125 000 million tons for what today would be described as Technically and Economically Recoverable Reserves, Clerk[14] calculated that the UK had about 500 years' supply

> .... at the present (1915) yearly consumption of 250 million tons.

Clerk's somewhat optimistic predictions for technically recoverable reserves were subsequently modified during the 1970s[9,15,16] to 45,000 million tonnes, enough to support current rates of production (circa 125 million tonnes per annum) for over 300 years.

Coal continued to dominate the energy market for the next 35 years, with annual outputs between 200 and 250 million tonnes. By 1950 coal still supplied 89.6% of the UK primary energy consumption, but its relative share started to fall as shown in Table 4.5. By 1971 it had been overtaken by oil. Figure 2.7 on page 38, which includes the major fuels from 1950 to 1983, could also be examined.

Table 4.5

Relative share of coal and oil in UK consumption
Consumption in m.t.c.e.

Year	Coal		Petroleum		Total consumption
	Consumption	%	Consumption	%	
1950	204.3	89.6	22.9	10.0	228.1
1955	216.9	85.4	36.0	14.2	253.9
1960	198.6	73.7	68.1	25.3	269.4
1965	187.5	61.8	106.2	35.0	303.3
1970	156.9	46.6	150.0	44.6	336.7
1971	139.3	42.1	151.2	45.7	331.0

Data from reference 4.

The same period saw the industry contract from a productive manpower of nearly 700 000 in 1955, with a deep mined output of 211 million tonnes and 901 collieries, to less than 290,000 in 1971 with 111 million tonnes of deep mined output and 289 collieries[17].

Table 4.6 shows the changes which occured in the distribution of UK coal consumption over the two decades from 1962 to 1982. The first decade saw a decline in several of the coal industry's traditional markets. The gas supply industry and coal-fired rail transport virtually disappeared and both the industrial and domestic markets fell to about 40% of their 1962 levels.

Table 4.6

Distribution of UK coal consumption
(million tonnes)

Year	1962	1972	1982
Electricity Supply Industry(1)	62.1	66.7	80.2
Gas supply industry	22.8	0.6	-
Coke ovens	24.1	20.4	10.4
Other conversion industries(2)	2.6	4.6	2.3
Industry	31.3	11.7	7.1
Railways	6.3	0.1	-
Domestic, including miner's coal	34.3	14.5	8.5
Others	10.8	4.2	2.5
Total home consumption	194.3	122.8	111.0
Stocks at end of year	41.1.	33.3	52.4

(1) Includes railway and transport power stations in 1982
(2) Includes gas works in 1982
Data from references 4,5 and 18.

Plan for Coal

The rise in the world price of oil in 1973 significantly altered the prospect for the coal industry. At that time the National Coal Board was having discussions with the government on its proposals for a 'Plan for Coal'[19]. The British Labour Party, then in opposition, also noted that no firm plans had been adopted for the future of the British coal industry. Following the return of the Labour Government to office in 1974, the new Secretary of State for Energy, Eric Varley, held a series of meetings which resulted in a full-scale examination of the industry by the National Coal Board, the unions and Government[20,21]. They formed a tripartite group with the following terms of reference

> To consider and advise on the contribution which coal can best make to the country's energy requirements and the steps needed to secure that contribution

The tripartite group accepted that they should mainly cover the period to 1985 but could look beyond that if necessary. They noted that the industry's decline since the middle of the 1950s had been due to the era of cheap oil and the technological changes in areas such as the gas industry and railways, outlined above. Similar declines had also been experienced in other countries, notably France, Belgium, Japan and Germany.

Among the more detailed objectives, they wished to assess the means by which coal could hold its competitive position in the long term. They also wished to establish how the industry's development could best be planned with equal emphasis given to the needs and welfare of consumers, coal miners and the coal industry, and the nation as a whole. The pattern of energy supply in the UK during 1973, shown in Table 4.7., was used as the base case. The group were naturally quite unaware that further oil price rises were still to come, followed by a major world economic depression, and that 1973 would continue to be the year of greatest UK energy consumption over the next decade. Their first assumptions were that the growth in overall demand for energy might be in the order of 10% up to 1980 and a further 10% up to 1985.

Table 4.7

Pattern of energy supply, 1973
m.t.c.e

Coal	Petroleum	Natural Gas	Nuclear Electricity	Hydro Electricity
131.1	159.4	39.7	9.9	2.0

In addition another 30 m.t.c.e. of oil and gas were used for non-energy purposes such as the production of petro-chemicals.

Data from reference 20.

This meant that the potential demand for coal to the mid 1980s could range up to 150 million tonnes per annum. The group noted that in common with most extractive industries the NCB 'had to run fast to stay still' and that it appeared likely that a broad average of some 3 to 4 million tonnes of capacity a year was likely to be lost, mainly through exhaustion of mines and possibly also through exceptional mining difficulties[20]. The first step in the plan was therefore to stablize coal production.

Plans were put forward to augment deep mined production by

(i) 9 million tonnes per annum by extending the life of existing pits

(ii) 13 million tonnes per annum by increasing the output at existing pits by major new projects

(iii) 20 million tonnes per annum from new collieries, including up to 10 million tonnes from the Selby coalfield.

Opencast production was planned to increase from its 1973 level of some 10 million tonnes per annum to 15 million tonnes.

The two other major parts of the Plan for Coal proposals were exploration and research. An extensive programme of exploration, test boring and seismic surveys was already in hand by 1974 to identify mineable coal[15,21]. This is defined as coming from deposits less than 1.2 km deep with seams at least 600 mm thick. The coal must also have an 'in situ' ash content of less than 20%. Research was to be broadly based, with proposals for a national development programme on coal conversion processes, such as supercritical gas extraction and liquid solvent extraction and on improved mining techniques and equipment. A key assumption on which the Plan was based was that output per man shift could be raised by some 4% a year, through the better application of proven techniques and from the development of new ideas, equipment and systems. A major scheme for settling the claims for damages by pneumoconiosis sufferers was included, as well as negotiations for the implementation of a 'sound and effective' incentive scheme. These plans, particularly for the new deep mined capacity, needed a very substantial capital investment.

Plan for Coal was adopted as the broad strategy for the development of the coal industry in July 1974. Short-term fluctuations in the price and availability of competing fuels were not to be allowed to interfere with steady progress in the implementation of the Plan. It was accepted that coal must remain competitive in the light of long-term trends. A sum of £600 million, at 1974 prices, was agreed for the programme.

The Plan was reviewed during 1976[15] and carried forward to the year 2000, in a report popularly known as Plan 2000[22]. The analysis for the likely UK demand for coal in the year 2000 was within the ranges shown in Table 4.8, and it was considered prudent to aim for a level about halfway in these ranges, 170 million tonnes, of which 150 million tonnes would be from deep mines and 20 million tonnes from open cast.

Table 4.8

Estimated UK demand for coal in 2000 AD

	million tonnes
Power stations	75 - 95
Coke Ovens	20 - 25
Industry	30 - 50
Domestic/Commercial including Synthetic Natural Gas	10 - 30
Total	135 - 200

This figure was based on the Department of Energy's forecast that the need for primary fuel, which included some non-energy uses, was likely to grow from the 1975 figure of 335.8 m.t.c.e.[21] to between 500 and 650 m.t.c.e. or more by the year 2000. Six years later the revised Department of Energy figures, presented as Proof of Evidence for the Sizewell 'B' Public Inquiry, ranged from 328 to 461 m.t.c.e.[23]. This represented a reduction of about one third from the figures used in Plan 2000 and would clearly alter the time scale relating to some of the developments agreed in Plan for Coal and Plan 2000.

By early 1984 it could be said that the principles of these Plans for Coal had been shown to be soundly based and that the industry had a firm basis on which to improve its future competitiveness. It was still being affected by short term adverse trade fluctuations. For example, the import of cheap Polish coal into some Western Europe countries had damaged export possibilities from the very large stocks which had built up since 1979 to the record figure of nearly 60 million tonnes by October 1983. The main trends in the decade up to the year ended 31st March 1983 are shown in Table 4.9.

Table 4.9

	Total output million tonnes	Opencast	Stocks (2)	Number of collieries	Manpower (thousands)	Output per man year (tonnes)
1973/4	108.8(1)	9.0	19.1	259	252.0	390
1974/5	127.2	9.2	21.6	246	246.0	474
1975/6	125.8	10.4	29.9	241	247.1	462
1976/7	120.8	11.4	28.1	238	242.0	448
1977/8	120.9	13.6	29.8	231	240.5	441
1978/9	119.9	13.5	28.8	223	234.9	448
1979/80	123.3	13.0	27.7	219	232.5	470
1980/81	126.6	15.3	38.4	211	229.8	479
1981/82	124.3	14.3	43.5	200	218.5	497
1982/83	120.9	14.7	53.3	191	207.6	504

Data from reference 17.

(1) Affected by strike action
(2) At the end of the year

Output per man year is for underground NCB mines. Output figures will differ slightly from those given in UK Energy Statistics which are normally based on calendar years.

The reduction in manpower averaged just under 2% per annum. The target of 15 million tonnes per annum for opencast mining had been achieved. The new deep mined output by 1985 will probably be some 7 million tonnes instead of 20, with the balance made up by new major projects in existing mines[22].

The planned overall increase in productivity had not been achieved, largely due to the problems of some small uneconomic collieries which had not been among the 68 which had closed. This is the subject of deep controversy between management and unions. However productivity at new mines and at existing mines with new major projects was significantly higher than average. Output per manshift averaged 4.35 tonnes at new mines and 2.64 tonnes at mines with major projects compared with a national average of 2.44 tonnes[17].

Governments have displayed a record of continuing confidence in the future of the industry over the decade since 1974. In real money prices more had been invested in the industry by 1983 than had been originally suggested in 1974. The announcement in January 1984 of the decision to open up the Vale of Belvoir coalfield, is further evidence that the long term developments outlined in the 1970s[16,20,21] will continue to receive priority.

Research and Development

The wide range of activities can be broadly divided into three areas - exploration, mining methods and coal utilization research. Of these, coal utilization research has been covered in greater detail as it is widely believed that coal will become increasingly important as a source for liquid fuels and chemical feedstocks in the next century.

Exploration

The vast resources of coal shown in Table 4.3. could support a greatly expanded use throughout the world for many years. As a direct result of the oil price increase in 1973, increased exploration for technically and economically recoverable reserves had discovered a further 185 000 m.t.c.e. by 1977[9]. In the UK there was also a rapid response. In the year 1972-3 the National Coal Board drilled 45 boreholes and carried out 32 km of seismic survey. Towards the end of the 1970s the rate had increased to about 150 boreholes and 400 km of seismic survey. With coal stocks rising to record levels in the early 1980s and the discovery of substantial additional reserves, the pace of exploration fell, with more concentration on exploring unworked reserves at existing mines. Fully explored reserves, ready for operation, totalled 7000 million tonnes by 1981, sufficient for over fifty years at present rates of extraction. It was the declared aim of the NCB to have this level of explored capacity always in hand.

Mining Methods

In the UK most of the underground operations are already mechanized and the emphasis has moved towards maintaining a high performance and reliability through improved control technology and monotoring techniques. For example, machine condition monitoring equipment allows the engineers to commission a coal face and monitor the face machines' condition and performance while operating. Desk top computers are used to keep records of machine health and maintenance costs[17]. An increasing number of different computer monitoring and controlling systems, known as MINOS (Mine Operating Systems), are reported each year. These cover coal clearance, ventilation, coal face coal preparation and fixed plant[22]. Longer term storage of face information, as well as on-line daily operational information had been installed in 22 collieries by 1983[17] as part of computer-based management information systems.

Fluidized-bed Combustion

The technology for fluidization, or causing small solid particles to behave as if they were a liquid, is well known and has been used in various applications for many years. It was first used as a method for firing boilers in the early 1960s. Fluidized-bed combustion requires the effective mixing and contact of the bed particles with the combustion gases to provide the optimum conditions for good combustion and heat transfer to the heating coils. This can be achieved by the design of the air distributor and solids feeding system[24]. Among the advantages attributed to the various fluidized-bed combustion systems are

(i) the ability to handle a wide range of fuels including lower grades of coal

(ii) sulphur dioxide can be absorbed in situ, reducing the level of atmospheric pollution by a factor of five or more compared with pulverized coal firing
(iii) nitrogen oxide emissions also appear to be reduced
(iv) that the capital and operating costs are lower.

The new technologies being developed by the National Coal Board should result in both improved power generation efficiences and a reduction in noxious emissions.[25] These use combined cycles with separate gas turbines and steam turbines. In the pressurized fluidized-bed cycle shown in Figure 4.3 the coal is burnt in a shallow fluidized bed of sand or ash and heats the pressurized air to form a gas to drive a gas turbine.

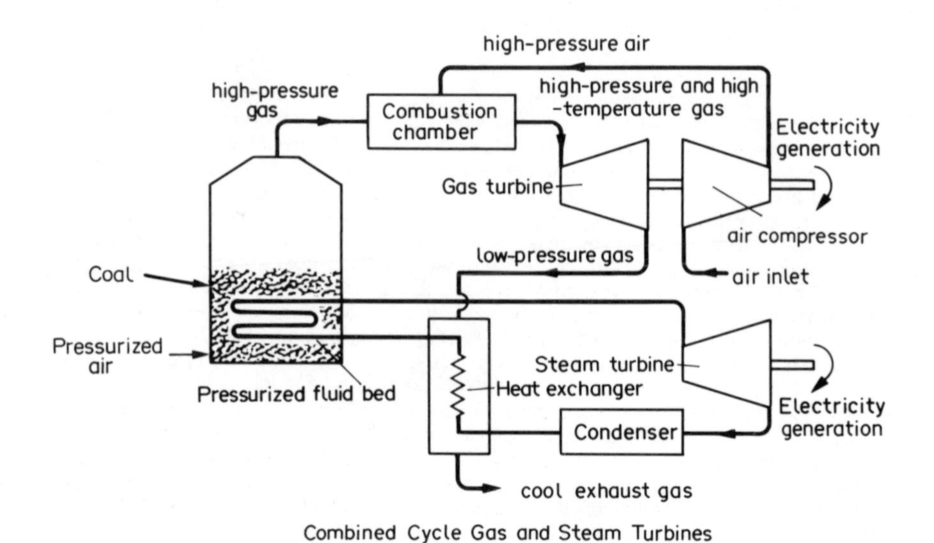

Figure 4.3

Another fluidized system, known as the gasification combined cycle converts coal into gas with the removal of ash particles and sulphur compounds. The clean gas is then burned in a gas turbine producing hot exhaust gases. The steam turbine cycle receives heat in two stages, the first stage heating being provided by the hot exhaust gases and the second stage during the combustion of the coal. Both these technologies are expected to be commercialy available by the end of the century.

Liquid Fuels from Coal

Methods for the production of liquids from coal have been known and practised for many years. For example the steel industry has produced liquids from coke ovens for over a hundred years. These liquids have been used as the primary raw materials for the coal tar and chemical industries. The main approaches to the production of liquid fuels from coal may be classified as follows[26]

(i) **Pyrolysis.** In all pyrolysis processes coal is heated in the absence of air. This breaks down the coal molecules, forming ethane and methane, leaving char behind. The tar products are primarily aromatic. Crude benzoil refining began in the nineteenth century and subsequently pure chemicals were obtained. Probably the best known modern industrial pyrolyzer is the Lurgi-Ruhrgas process[27].

(ii) **Direct Hydrogenation.** A reducing gas, such as high-purity hydrogen, reacts with the coal at high temperatures and pressures in the presence of a catalyst. The products are also primarily aromatic. The process was developed commercially in Germany and the UK from the mid 1920s. By 1935 the I.C.I. plant at Billingham had a capacity of 100 000 tonnes per annum of liquid fuels[26].

(iii) **Solvent Extraction.** In this process, finely crushed and dried coal is treated with a hot liquid solvent derived from coal. After processing it is possible to generate hydrogen as well as a range of hydrocarbons which can be upgraded to petrol or diesel fuel. The heavy residual oils be recycled as the solvent oil[22].

(iv) **Indirect Liquefaction.** This is sometimes called the synthesis process, as the feed coal is initially completely gasified to produce a synthesis gas. The gas is purified to remove sulphur compounds and solid particulates. It then reacts at high temperatures and pressures in the presence of catalysts to produce liquid products. The best known process was first developed in Germany in the 1920s by Franz Fischer and Hans Tropsch and the first Fischer-Tropsch plant was built there in 1936. The only large commercial plants operating by the early 1980s were the Sasol Plants in South Africa. Sasol I commenced operating in 1955 and produces a wide range of products, including petrochemicals and fertilizers. When Sasol III is completed in mid-1980s, the three plants will use some 30 million tonnes of coal per annum to produce petrol, diesel fuel, liquified petroleum gases and kerosene as well as other alcohols and chemicals[26].

Another process, which is being examined in the UK, is the supercritical gas-extraction process. High pressure gas is used to dissolve the coal. At room temperature the extract is a glass solid which can be processed to yield liquid fuels and chemical feedstocks. An interesting feature is that the extract contains 7% by weight of hydrogen compared with approximately 5% by weight in the original coal[22].

Coal-Oil Mixtures

Coal-oil mixtures, coal-in-oil fuels, oil-coal suspensions are just a few of the terms used for many years to describe a mixture of solid particles of coal and a liquid hydrocarbon fuel, often a crude oil. These mixtures have often been used when supplies of petroleum were limited, particularly during the Second World War[25]. A number of countries are again developing new boiler designs as a result of the relatively high oil prices. The main advantages of coal-oil mixtures are that coal can be used as a substitute for oil, with improved combustion, handling, metering and control compared with burning the coal directly.

Health, Safety and the Environment

The main issues of public concern vary from one country to another and within each country. For the miners there are occupational health hazards. The communities surrounding the mines will be concerned about land reclamation following surface mining and about the effects of subsidence from underground mining. Coal waste and coal preparation plants can create acid drainage. The combustion of coal creates noxious emissions, such as sulphur dioxide, and particulates, while carbon dioxide is widely believed to have an important influence on world climate. The safe disposal of large quantities of ashes from coal-fired power stations can create problems. Some of these issues are examined in detail below.

Occupational Health Hazards

The major occupational health hazard associated with underground mining has been black lung disease, or pneumoconiosis, caused by the breathing of coal dust. The two worst safety hazards have been from gas explosions and flooding. Dust and gas levels have been greatly reduced in recent years by improved ventilation and filtration systems. Dust has also been suppressed by water spraying. Nevertheless visitors to deep underground mines are often conscious that they are breathing some coal dust and can experience discomfort[13]. In the UK the NCB has had a good record in reducing the risk of pneumoconiosis and they carry out a wide variety of health screening activities[17]. Mining must still be regarded as a relatively dangerous occupation, although there has been a steady decline in the number of fatal accidents in the UK in the past thirty years. In the 1950s the chance of a fatal accident occuring to a worker during the year was about 1 in 1500. By 1982/83 this had fallen to about 1 in 5000.

The causes of the major catastrophic mine accidents of the past are now much better understood. Minor accidents have also been reduced through improved training and safety programmes[9]. Where the best current safety practices are followed, the accident and illness rates experienced with coal mining are comparable with construction work or many sectors of heavy industry.

Acid Precipitation

Acid precipitation, or 'acid rain', is largely caused by the interaction of sulphur dioxide and nitrogen oxide emissions with water in the atmosphere. The emissions are produced from the combustion of coal and oil. Power plants, smelters, other industrial sources and motor vehicles cause these emissions. The gases can be dispersed by tall chimneys which prevent high local concentrations at ground level. But this approach, which has been adopted in many countries including the UK, may allow increases in the acidity of rain up to hundreds of kilometres from the source. This can accelerate the leaching of toxic metals, such as aluminium and mercury, into groundwater and watercourses. There is evidence of damage to forests and fish stocks, particularly in areas with acid soils[25,28]. The problem can only be tackled at the source. Several countries have introduced legislation controlling the emissions of sulphur oxides from fossil-fuelled power stations but there are other complications in control. The largest single source of nitrogen oxides is motor vehicle emissions. Acid precipitation in some countries, such as Norway, Sweden or Eastern Canada, results from emissions in other countries where there is a reluctance to impose controls. International collaboration is the only way to approach these issues. In the UK the National Coal Board and the CEGB have funded a major international collaborative research programme by the Royal Society and the Royal Swedish and Norwegian Acadamies of Science into the causes of acidification of surface waters in affected areas of Norway and Sweden[29].

The coal industry is also examining the possibility of reducing sulphur oxide emissions by reducing the sulphur level of the coal or by modifying coal-burning techniques. Flue gas desulphurization was used in the UK up to the 1960s, when the tall chimney policy was adopted. Other countries have continued with this technique, but it has the major disadvantage of needing associated solid sulphur disposal. In the short term the adoption of either process will reduce generating efficiency and increase power station costs[25].

Carbon Dioxide Levels in the Atmosphere

The theory that increases in CO_2 levels in the atmosphere will warm the earth, the 'greenhouse theory' was first proposed at the end of the nineteenth century. The theory states that certain gases in the atmosphere allow the ultraviolet and visible radiation from the sun to pass through and warm the earth, but then absorb the longer wavelength infra-red radiation emitted from the earth's surface. This creates a thermal barrier around the earth, raising the earth's temperature. One estimate[30] suggests that the average global temperature of 288K is approximately 35K warmer that it might have been without the greenhouse layer. The main greenhouse gas is CO_2, but other gases such as methane, nitrous oxide and chlorofluorcarbons, and water vapour have similar effects. Increases in the levels of any of these gases could alter temperatures on earth.

As coal consists mainly of carbon, it releases considerably more CO_2 than oil or natural gas for the same amount of heat supplied. From 1860 to the early 1980s the concentration of CO_2 has grown from about 280 parts per million (ppm) to 339 ppm[31]. Any major increase in CO_2 emissions caused by continuing growth in the use of coal would cause the level in the atmosphere to build up steadily. The oceans would first absorb and distribute the additional heat produced by the greenhouse effect, so global temperature rises would lag behind CO_2 increases perhaps by several decades.

Mathematical modelling of global climate and temperature is very complex, but recently more sophisticated models have shown a much better fit when the effects of volcanoes and variations in incoming solar radiation are taken into account as shown in Figure 4.4.[32].

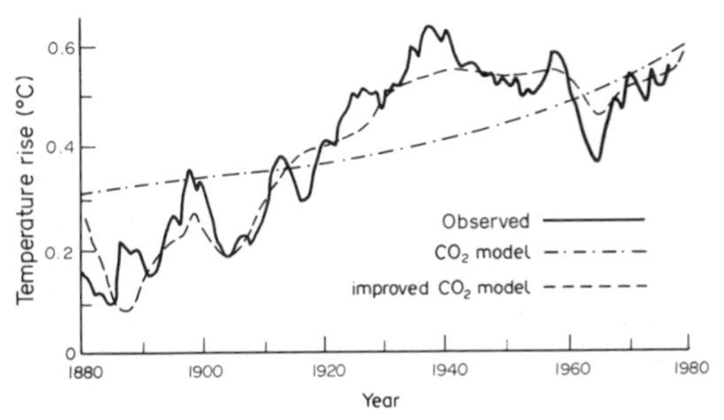

Figure 4.4

By 1983 a report from the Environmental Protection Agency[33] stated that while there was considerable uncertainty as to the rate and ultimate size of global temperature rise, a 2°C increase could occur by the middle of the next century and a 5°C increase by 2100. This could be accompanied by 'dramatic changes in precipitation and storm patterns and a rise in global average sea level.' They concluded that the risks were high in pursuing a 'wait and see' attitude on one hand, or in acting impulsively on the other.

Peat

Peat can be regarded in geological terms as 'young coal' and is a heterogeneous mixture of partially decomposed organic matter and inorganic minerals that have accumulated in a water-saturated environment[34]. It has been used on a small scale in many countries for over two thousand years and in limited industrial use since the beginning of this century. The major problem in its use as a fuel is its very high moisture content, which in its natural state is often over 90%. This has to be reduced to 55% or less before it can be burnt. The world production of fuel peat in 1980 was very small compared with all other major resources, representing the equivalent of 25 million tonnes of oil, or less than 0.5% of the world production from the fossil fuels.

There are four basic forms of fuel peat, depending on the required use and production method. The traditional method is to dig the peat manually from the bogs. Hand-cut peat is normally shaped in the form of a retangular block, typically 15 x 20 x 30 centimetres and is used in individual homes for heating and cooking. Sod peat is mechanically cut and then compressed during the production stage to a cubical or cylindrical form, slightly smaller than the traditional hand-cut block. It is also used in individual homes and for small scale commercial use. Milled peat is a heterogeneous mixture of small loose peat particles mechanically cut from the surface of the bog. Typical cutting depths are about two centimetres and the particle sizes range from three to eight millimetres. It is used mainly in power stations and in large industrial boilers. Peat briquettes are usually produced from milled peat which is thermally dried to 10%-20% moisture content and then compressed into briquettes. In Ireland these are sold in 10 kg packages, approximately 15 x 20 x 45 centimetres, which can be split into smaller blocks for use in individual homes. The main characteristics of the different types of fuel peat are given in Table 4.10.

Table 4.10

Characteristics of Fuel Peat

Type	Heating Value MJ kg^{-1}	Moisture Content %	Density kg m^{-3}
Hand-cut	11-15	25-40	200-400
Sod	11-14	30-40	300-400
Milled	8-11	40-55	300-400
Briquette	17-18	15	700-800

Data from reference 34

Estimates for the total world resources of peat are very approximate and are broadly equivalent to the proven reserves of petroleum or natural gas, in the order of 100 billion tonnes of oil equivalent. Over a half of all the world's peat resources are estimated to be in the USSR and about a third in Canada[34]. There is practically no data on tropical and subtropical peat. Other countries with substantial peat deposits include the United States, Indonesia, Finland, Sweden, the People's Republic of China, Norway, Malaysia, the UK, Poland and Ireland. Of these, only three countries can be regarded as major peat producers, the USSR, Ireland and Finland, who produced 80, 6 and 3 million tonnes of fuel peat respectively in 1980 out of an estimated world total of 90 million tonnes (on a 40% moisture content basis). The largest use of peat was not for fuel but in horticultural applications where again the USSR led world production with 120 million tonnes out of an estimated world total of 130 million tonnes[34].

Peat will never be a major world energy resource, but can have considerable local importance in countries with little or no resources of petroleum, natural gas or coal. It is only significant at national level in Ireland, where it supplies some 20% of primary energy consumption. In the other two peat producing countries it accounts for well under 5% of their primary energy use.

Summary

Most authorities agree that coal is capable of meeting a larger proportion of the world demand for energy than at present. Economically recoverable reserves are several times those of oil and gas and are clearly capable of meeting an increased demand for at least several centuries. The large scale conversion of coal to liquid and gaseous fuels could become necessary as reserves of the other main fossil fuels, oil and gas, decline. The timing of this change of emphasis is uncertain, but could probably occur during the next fifty years. The direct combustion of coal has certain environmental consequences. Some can be met by legislation, but the effect of global warming through the increased CO_2 levels in the atmosphere may pose very severe problems. It is highly unlikely that the optimistic suggestion in the 1980 World Coal Study of an increase in coal production by a factor of $2\frac{1}{2}$ or 3 during the twenty year period to the end of the century could ever be achieved. This would need annual growth rates at least twice as high as the levels experienced in the past decade. In the UK the Government believes that the coal industry has an important and vital future, although concern was expressed early in 1984 that the 4% per annum increase in productivity agreed in the 1974 Plan for Coal had only been achieved once, in 1983, and that the least economic areas of coal production had been closed down at a rate of about half that originally planned[35].

References

1. Foley, Gerald, The Energy Question, Second Edition, Penguin Books, 1981.

2. Williams, D.G., Coal Cleaning, in Coal Handbook, Marcel Dekker Inc., New York and Basel, 1981.

3. Coal Resources, World Energy Conference, London, 1978.

4. Digest of Energy Statistics, HMSO, London, 1971.

5. Digest of United Kingdom Energy Statistics, HMSO, London, 1980.

6. Dorf, Richard C., Energy Resources and Policy, Addison-Wesley, 1978.

7. Energy in the Developing Countries, World Bank, Washington, 1980.

8. Coal Worldwide, Petroconsultants Ltd., Dublin, Ireland, 1980.

9. Coal - Bridge to the Future, Report of the World Coal Study (WOCOL), Ballinger Publishing Co., Cambridge, Massachusetts, 1980.

10. Future Coal Prospects: Country and Regional Assessments, Report of the World Coal Study (WOCOL), the second and final report, Bollinger Publishing Co., Cambridge, Massachusetts, 1980.

11. Koch, J., Solid Fossil Fuels, in Survey of Energy Resources 1980, London, World Energy Conference 1980.

12. Holton, J.E.B., World Coal, in Assessment of Energy Resources, Report Number 9, The Watt Committee on Energy, London, 1981.

13. McVeigh, J.C., This section is based on personal experience at the coalface at Betteshanger, NCB South Midlands Area, 1981.

14. Dugald, Clerk, The world's supplies of fuel and motive power, 1915 Thomas Hawksley lecture, Institution of Mechanical Engineers, London, Proc. I. Mech E., 591-625, October 1915.

15. Ezra, Derek, Coal, in The Energy Debate, Benn Technical Books, Croydon, Surrey, 1983.

16. Coal for the Future, Progress with 'Plan for Coal' and prospects to the year 2000. Department of Energy, London, 1977.

17. Report and Accounts 1982/83, National Coal Board, London 1983.

18. Energy Trends, Department of Energy, London, December 1983.

19. Plan for Coal, The National Coal Board, London, 1973.

20. Coal Industry Examination, Interim Report, Department of Energy, London, June 1974.

21. Coal Industry Examination, Final Report, Department of Energy, London, 1974.

22. Ezra, Derek, A Review Plan for Coal, The Mining Engineer, January 1981.

23. Proof of Evidence for the Sizewell 'B' Public Inquiry, Department of Energy, London, October 1982.

24. Meyers, Robert A., Ed. Coal Handbook, Marcel Dekker Inc., New York and Basel, 1981.

25. Davies D.D., Future Markets for Coal, Energy World, 106, 4-11, 1983.

26. O'Hara, J.B., Liquid Fuels from Coal, in Coal Handbook, Marcel Dekker Inc., New York and Basel, 1981.

27. Marnell, P., in Synthetic Fuels Processing, Marcel Dekker Inc., New York, 1977.

28. Schrecker, Ted., The Conserver Society Revisited, Science Council of Canada, Ontario, May 1983.

29. £5 million 'Acid Rain' Research, Press Release 1817, National Coal Board, London, September 1983.

30. Chamberlain, J., et. al., Climatic Effects of Minor Atmospheric Constituents, in Carbon Dioxide Review, 1982, Clarendon Press, New York, 1982.

31. Keeling, C.D., et.al., Measurements of the Concentrations of Carbon Dioxide at Mauna Loa Observatory, Hawai; Ibid.

32. Hansen, J. et. al., Climate Impacts of Increasing Atmospheric CO_2, Science, 213, 957, 1981.

33. Seidal, S., and Keyes, D., Can we delay a greenhouse warning? US Environmental Protection Agency, Washington, September 1983.

34. Report on the use of peat for energy, United Nations Conference on New and Renewable Sources of Energy, Nairobi, 1981.

35. Walker, Peter, Approval for Ashfordby Mine in North East Leicestershire, Press Notice 2, Department of Energy, London, 9th January 1984.

Further Reading

For a general background to both resources and prospects for the future the World Coal Study, Coal-Bridge to the Future, Ballinger Publishing Co., Cambridge, Massachusetts, 1980, presents an optimistic picture. Coal Handbook, edited by Robert Meyers, Marcel Dekker Inc., New York and Basel, 1981, is a comprehensive and authoritative source book which covers the whole field of coal technology in considerable depth. Gerald Foley and Ariane van Buren, of the International Institute for the Environment and Development, London, discuss problems in the Third World in Coal Substitution and other approaches to easing the pressure on world fuel resources, prepared for the United Nations Conference on New and Renewable Sources of Energy, November 1980. Among several publications on CO_2 problems, the United States Environmental Protection Agency's Can We Delay a Greenhouse Warning?, by Stephen Seidel and Dale Keys, Washington, September 1983, is an invaluable reference source.

In the UK Sir Derek Ezra, former Chairman of the NCB, has published a number of books and articles including Coal and Energy, Ernest Benn Ltd., London and Tonbridge, 1978. The Vale of Belvoir Inquiry is covered by one of the main witnesses for those opposing the National Coal Board's plans, Professor Gerald Manners, in Coal in Britain. An Uncertain Future, George Allen and Unwin Ltd., London, 1981. Many useful booklets and general policy statements are always available from the Press Office of the National Coal Board, Hobart House, Grosvenor Place, London.

Exercises

1. In the World Coal Study the summary stated: 'The present knowledge of possible carbon dioxide effects on climate does not justify delaying the expansion of coal use'. Discuss this statement with reference to today's knowledge.

2. The United Nations Economic Commission for Europe has established a 'Long-range transboundary air pollution convention' to investigate the best policies for controlling acid precipitation. Examine the present state of this and other investigations into the problems created by dilute nitric and sulphuric acids in the atmosphere.

3. What are the main features of the coal mining industry in the United States, the USSR and the People's Republic of China?

4. The South African Sasol plants could serve as the models for future major coal processing systems. Examine their present developments, paying particular attention to the economics of the major processes.

5. By selecting no more than six time intervals sketch the curves for cumulative discoveries (technically and economically recoverable) and cumulative production for coal, in the world and in the UK. Explain why the logistic curve approach no longer appears to be appropriate for the UK. What are the limitations to this approach on a world basis?

6. Coal goes through a number of stages from the initial extraction to the final supply of electricity to the home. List these stages and, by determining the amount of energy spent during each stage, find what percentage of the energy originally available in the coal is supplied as useful 'end-use'energy.

7. By October 1983 the National Coal Board in the UK was producing some 8 to 10 million tonnes of coal a year above the demand. What steps has the industry taken since then to solve this problem while safeguarding the interests of the workers?

8. The Selby mine in Yorkshire is the largest project ever undertaken by the British coal industry. It will become the most extensive deep-mine complex in the world. What are its main features? It is planned to reach full production by 1987/8. What stage has the development reached?

9. In 1978 the National Coal Board applied for planning permission for both underground and surface developments in the North East Leicestershire area - the Vale of Belvoir. What were the main reasons put forward by the Coal Board in favour of this development? What were the main objections? What is the present position?

5. GAS

NATURAL GAS

The development of natural gas as one of the three major sources of primary energy in the world was briefly discussed in Chapter 2. The term natural gas is used to describe the mixtures of hydrocarbon and non-hydrocarbon gases found in subsurface sedimentary rock reservoirs, sometimes associated with petroleum[1]. The majority of the world's natural gas in found quite independently from an oil field and is known as 'non-associated' gas to distinguish it from 'associated' gas. Associated gas can be found in a layer or 'gas cap' above the oil in a reservoir, or as gas present and dissolved under pressure in the crude oil. The gas/oil ratio is usually expressed in cubic feet of gas measured at normal temperature and pressure (15°C and 760 mm of mercury) per barrel of oil, and can vary from less than 100 to several thousand cubic feet per barrel[1]. The origins of natural gas are the subject of controversy. Associated gas is believed to have the same marine origins as petroleum, but non-associated gas may be derived from vegetable matter possibly from the natural degasification of coal at great depths[2]. The liquids and gases in the same associated field may not have had a common origin, as the natural gas could have migrated from much deeper rock layers.

The composition of natural gas fields varies considerably. The main constituent gas is methane (CH_4) which often forms between 85 and 95 per cent of the total. The other hydrocarbon gases in natural gas mixtures are all members of the paraffin series, with the general formula C_nH_{2n+2}[1]. After methane (n = 1), the next four constituents are ethane, propane, butane and pentane, with values of n from 2 to 5 respectively. 'Wet' natural gas contains varying quantities of the heavier hydrocarbon compounds, such as propane, butane and pentane. They are known collectively, when liquidified, as natural gas liquid (NGL) or sometimes as condensate[2]. They may occur as liquids or exist with oil in one indistinguishable phase in some high pressure reservoirs and are formed only with a sudden release of pressure. The gas is 'wet' if the extractable liquid hydrocarbon content is more than one litre per 21 cubic metres of gas (0.3 gallons per 1000 cubic feet) and 'dry' if it is less than one litre per 63 cubic metres of gas[1]. A 'lean' gas lies between these limits. Table 5.1. shows typical analyses of wet, dry and condensate hydrocarbons.

Table 5.1

Typical Analyses of Natural Gas Hydrocarbons

Hydrocarbon	Condensate%	'Wet' Gas %	'Dry' Gas %
Methane	5.1	84.6	96.00
Ethane	3.5	6.4	2.00
Propane	9.2	5.3	0.60
Iso-butane	5.0	1.2	0.18
N-butane	7.4	1.4	0.12
Iso-pentane$^+$	5.8	0.4	0.14
N-pentane$^+$	4.3	0.2	0.06
Hexanes$^+$	11.8	0.4	0.10
Heptanes$^+$	47.9	0.1	0.80

Data from reference 3

+ Liquid at 15°C and atmospheric pressure

The hydrocarbons are separated from the wet gas to obtain the dry natural gas which is sold to consumers. Ethane is used in the petrochemical industry and both propane and butane are liquified and sold as Liquified Petroleum Gas (LPG).

Variable quantities of several non-combustible gases can also be present. Nitrogen, hydrogen sulphide and carbon dioxide can occur in substantial percentages, but only small amounts of argon or helium are present as indicated in Table 5.2. Hydrogen sulphide (H_2S) is actively poisonous and corrosive and a natural gas contaminated with it is known as a 'sour' gas. A 'sweet' gas is generally one in which there is no detectable odour present, typically with less than one part in a million of hydrogen sulphide. Large amounts of the inert gases such as nitrogen or helium lower the calorific value of the natural gas.

Table 5.2

Typical Occurances of Non-Hydrocarbons in Natural Gases

Gas	Typical %	Maximum Recorded %
Carbon Dioxide	0-5	98
Helium	0-0.5	8
Hydrogen Sulphide	0-5	63
Nitrogen	0-10	86
Argon	less than 0.1% of the helium content	
Radon, Krypton, Xenon	traces only	

Data from reference 1

Development of the industry

The foundations of the modern natural gas industry were laid in Fredonia, in New York State when natural gas was piped through wooden tubes or hollow logs from a shallow well used for street lighting in 1821[1]. The first recognized natural gas pipeline distribution system was completed in 1883, taking gas to Pittsburgh from a field 22 km away[1,4]. At the same time the development of the Baku oil field near the Caspian

Sea with the associated natural gas had also commenced[1]. However, production and exploitation of natural gas until the middle of the twentieth century was largely confined to the United States. As late as 1950 over 85% of the world's natural gas was produced and consumed in the United States, with only two other countries, Venezuela and the USSR with 8% and 4% respectively, recognized as producers[5]. Meanwhile in Europe the comparatively modest efforts to find any significant natural gas fields met with no success until the discovery of the Dutch Slochteren field by a joint Esso-Shell survey team on the 14th August 1959. Further exploration revealed that this was one of the world's largest gas fields and a discovery which had a major impact on the energy economy of northern Europe. The geological features indicated that the gas was contained in porous sandstone trapped under an impermeable layer of salt. This gas was thought to have originated from deep coal layers and it was considered that any region with a thick salt layer above deep coal had the potential to be a natural gas field[6]. Many of these regions proved to be in the North Sea and all the countries with coastlines on the North Sea - the Netherlands, the United Kingdom, Belgium, Germany, Denmark and Norway - entered into agreements defining the extent of their sovereignity. On the 17th September 1965, just over six years after the first Dutch discovery, the British Petroleum Company found gas some 72 km east of the Humber. The subsequent growth of the natural gas industry in the UK was described in Chapter 2.

From 1950 the dominant position of the United States as the world's major producer and consumer started to decline. There were two reasons for this. Although the annual growth in production averaged about 6.4% for the next two decades, the size of new discoveries began to decrease compared with consumption. In 1968 consumption exceeded additions to the reserves for the first time and a detailed examination by Hubbert[7], using logistic curves, predicted that gas production in the United States would peak between 1978 and 1980 at an annual production level of 21 to 25 Tcf. The second reason was the emergence of the USSR as a major producer with much greater potential resources than any other country in the world. The relative position of the two countries from 1965 to 1982 is shown in Table 5.3.

Table 5.3

Annual Production

	USA	USSR	Combined percentage of world production
	Trillion cubic feet Tcf		
1965	16.04	4.51	81.0
1970	21.92	6.99	77.3
1975	19.24	10.22	66.4
1980	19.56	16.15	67.6
1982	17.69	17.69	65.9

Data from references 8, 9 and 10

Production in the United States actually peaked in 1972, a few years earlier than predicted by Hubbert. While the growth in production in the USSR has slowed over the past decade, it is still very high at some 8% per annum.

Consumption, Production and Proved Reserves

World gas consumption during the period from 1965 to 1982 is shown (in million tonnes of oil equivalent) in Figure 5.1.

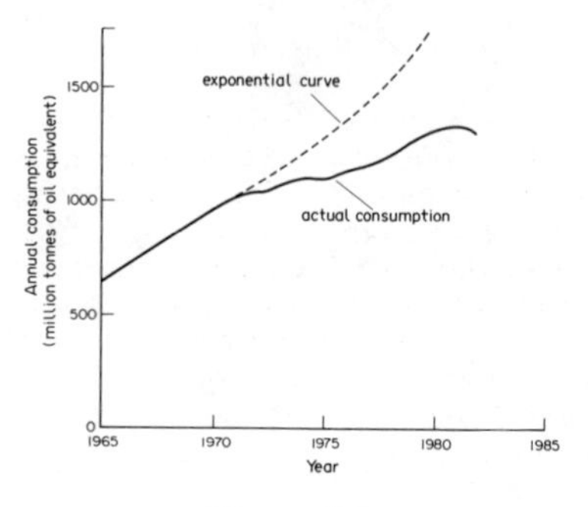

Figure 5.1

The general pattern of a growth rate of 7% per annum up to 1972 is similar to that experienced with the other major fuels. The limitations of assuming that this 7% per annum growth rate could continue into the 1980s are shown when actual consumption is compared with the theoretical exponential growth curve. Actual growth rates averaged about 2.6% over the next decade, with the effects of the oil price rises in 1973 and 1979 clearly marked. World consumption and production during 1982 in selected countries and regions is shown in Table 5.4. The main difference between these figures and those for oil is that 90% of all natural gas is used in the countries in which it is produced.

Table 5.4

World natural gas consumption and production 1982

	Consumption	Tcf	Production	
Netherlands	1.07		2.14	
Norway	Nil		0.82	
UK	1.63		1.27	
TOTAL WESTERN EUROPE		6.83		5.71
Canada	1.69		2.47	
United States	18.15		17.69	
TOTAL WESTERN HEMISPHERE		22.19		22.85
ASIA-PACIFIC		2.20		2.32
AFRICA		0.71		1.19
MIDDLE EAST		1.49		1.73
China	0.37		0.39	
USSR	14.90		17.69	
TOTAL CENTRALLY PLANNED ECONOMIES		18.02		19.91
TOTAL WORLD		51.44		53.71

Derived data from reference 8

The world's natural gas reserves have been less thoroughly explored than oil reserves and one leading analyst[11] commented in 1983 that this fact alone 'justifies greater optimism in regard to future discoveries'. During the early 1950s some associated natural gas was considered to be 'a nuisance product' of oil production[12]. This has meant that any subsequent estimates of cumulative gas production cannot be accurate, as large quantities of associated gas were flared and wasted in the early days of the oil industry. In the United States legislation prohibiting flaring was eventually introduced as there were alternative uses for the gas - a well established petrochemical industry and a major pipleine distribution network. The term 'production' in the early 1960s also included estimates for the flared gas and Tiratsoo's suggestion[1] that only 80% of the estimated total world natural gas production in 1965 should be regarded as 'consumed', gave a figure for world natural gas consumption of 669 m.t.o.e., which is very close to the widely quoted figure of about 647 m.t.o.e. If actual consumption figures are assumed to be 80% of total production, then taking Tiratsoo's estimate for cumulative production up to 1965 as 340 Tcf, the cumulative production figure by the end of 1982 would be 1265 Tcf. This is in good agreement with Halbouty's figure of 1313 Tcf given at the World Petroleum Congress[12] in 1983. The curves for cumulative production, proved reserves and cumulative discoveries are shown in Figure 5.2.

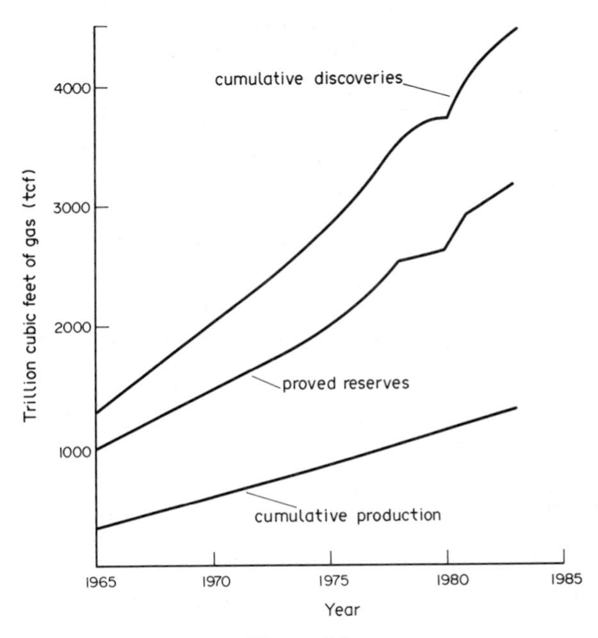

Figure 5.2

These can be compared with the curves in Figure 3.3. for oil and the theoretical Figure 1.6. By the beginning of 1984 natural gas was clearly at an earlier stage in its life cycle compared with oil, as proved gas reserves were still rising at a greater rate than cumulative production. This was confirmed[13] by a 5.4% rise in total proved reserves during 1983, continuing the trends since 1961 as shown in Table 5.5. Growth in reserves during the early 1960s was about 8% per annum.

Table 5.5

World proved reserves of natural gas
Trillion cubic feet (Tcf)

Year	Tcf
1961	721.5
1965	998.7
1974	1855.0
1978	2520.5
1979	2576.8
1980	2639.6
1981	2911.9
1982	3033.5
1983	3199.9

Data from references 8,9,13,14 and 15

The distribution of proved reserves between the various countries and regions is given in Table 5.6 for all countries with reserves greater than 20 Tcf.

Table 5.6

World proved natural gas reserves on 1-1-84
and reserves/1982 production ratios

	Reserves	Tcf	% of world+ reserves	Reserves/+ production ratio
Indonesia	30.2		0.94	44.1
Malaysia	48.0		1.50	-
TOTAL ASIA-PACIFIC		156.2		
Netherlands	50.1		1.57	23.4
Norway	58.8		1.84	71.7
UK	25.1		0.78	19.8
TOTAL WESTERN EUROPE		157.3		
Abu Dhabi	20.5		0.64	44.3
Iran	480.0		15.00	1882
Iraq	29.0		0.91	-
Kuwait	31.0		0.97	182.3
Qatar	62.0		1.94	-
Saudi Arabia	121.0		3.78	235.0
TOTAL MIDDLE EAST		775.0		
Algeria	110.2		3.44	122.4
Libya	21.4		0.67	152.9
Nigeria	34.8		1.09	870
TOTAL AFRICA		189.6		
Argentina	24.4		0.76	71.8
Canada	90.5		2.83	36.6
Mexico	75.4		2.36	58.7
Venezuela	54.5		1.70	94.8
United States	198.0		6.19	11.2
TOTAL WESTERN HEMISPHERE		474.9		
China	30.0		0.94	76.5
USSR	1400.0		43.75	79.1
TOTAL CENTRALLY PLANNED ECONOMIES	1446.8			
TOTAL WORLD		3199.9		

Data from references 8 and 13. All values are rounded.
+ Author's derived data

The table emphasises the declining position of the United States with less than 7% of the world's reserves and a reserves/production ratio of only 11.2 compared with the USSR's figure of nearly 44% of world reserves and a reserves/production ratio of 79.1. Reserves in several of the Middle East oil producing countries where only limited exploration for gas has taken place could be very much larger. Iran, with 15% of proved reserves, had a reserves/production figure of nearly 2000 and has the potential to be a major producer until well into the next century.

Ultimately Recoverable Resources

Estimates of the world's ultimately recoverable resources of natural gas are subject to greater uncertainties than those for the proved reserves which were outlined in the previous paragraphs. For example Ion[14] draws attention to the controversy over whether some of the associated gas should be regarded as oil or gas and also whether some known associated gas reserves should be included because their utilization depends on the operating decisions relating to the crude oil. In some associated fields, for example in the North Sea, gas is reinjected to maintain pressure for oil extraction. Such gas would only become available when the oil field is exhausted. Probably the largest figure to be quoted in the literature was suggested by the United States Geological Survey in the 1950s, when they concluded that some 19,100 Tcf may eventually be discovered. A full explanation of all their various arbitrary assumptions was given by Tiratsoo[1] who also pointed out that subsequent estimates, such as the 1956 United States Mackinney Report with 3,700 Tcf and the 1962 World Power Conference with 2,500 Tcf, were considerably lower. However during the 1960s and 1970s a range of estimates from 9,500 to 12,000 Tcf were made[14], although other more pessimistic estimates were perhaps only a quarter of these figures. The recent confirmation of the continuing upward revision in the proved reserves shown in Table 5.4. was just one of the factors which resulted in the figure of 9,600 Tcf being presented to the 1983 World Petroleum Congress[12]. This estimate was based on studies of the world's sedimentary basins with the latest engineering, geological and statistical data. Two important assumptions were that offshore exploration would not exceed depths of 2500m and that the overall recovery factor would be 72%. Although this figure of 7,600 Tcf must be regarded as a modal point between values perhaps 30% lower which could be regarded as highly probable to values somewhat greater which might be attainable, it puts the ultimately recoverable natural gas resources up to the same level as the modal values for oil shown in Chapter 3 in Figure 3.5.

International Trade

Although internationally traded gas increased from some 3% of total consumption in 1966 to 13% by the end of 1982, this was still a relatively small fraction compared with internationally traded oil which was 45% of total oil consumption in 1982[8].

Table 5.7

Natural Gas Pipeline and LNG Exports 1982

Country	Tcf	Percentage share
USSR	2.79	40
Netherlands	1.07	15
Norway	0.82	11.5
Canada	0.78	11
Other Pipeline exports	0.32	4.5
Indonesia+	0.46	6.5
Algeria+	0.35	5
Brunei+	0.25	3.5
Other LNG exports	0.21	3
TOTAL	7.05	100

+ Liquified natural gas (LNG)

Data derived from references 8 and 15

Exports of pipeline gas were dominated by the USSR and three other countries as shown in Table 5.7. Liquified natural gas is exported in special cryogenic ocean carriers and represented about one fifth of total world exports. LNG ocean carriers are normally used for longer distances while pipelines are considered more suitable for shorter distances. Long distance pipelines connect the USSR to both Eastern and Western European countries; the Netherlands exports to West Germany, France, Belgium, Luxembourg, Italy and Switzerland; Norway to the UK, West Germany, France, Belgium and the Netherlands while Canada exports to the United States. Trade in LNG is expected to grow substantially towards the end of the century[2] with some ten countries including the USSR and Iran considering new LNG liquification plants and terminals.

Other Factors influencing development

Natural gas has always been considered as a clean burning, easily controlled fuel with many other advantages. It has a high calorific value (about 37 MJ m^{-3}) and can be used for direct heating in both industrial and domestic applications. It can be used as a fuel for gas turbines or certain types of internal combustion engine and hence for generating electric power. It can be used as raw material for the petrochemical industry. The environmental impact of natural gas when burnt is negligible, apart from increasing the level of carbon dioxide in the atmosphere. The transmission pipelines are often buried underground, eliminating visual impact. Storage is not needed on the site where it will be used. A major disadvantage which has limited the use of natural gas in the past has been that the gas fields have been remote from their markets and that it is comparatively expensive to transport and distribute. From the late 1970s the higher levels of energy prices have increased the incentives for production and fields which were known but regarded as uneconomic were becoming more attractive[2]. Approximately 6% of world production was reinjected into oil reservoirs (and therefore available for future use) and flaring was estimated to be at about 9% of world production outside the USSR and the centrally planned economies.

GAS FROM COAL

In most industrial countries outside the United States the term 'gas industry' referred to the coal gas industry until the developments of world natural gas from the 1950s. The actual discovery of coal gas cannot be attributed to any particular individual but it was well known by the beginning of the eighteenth century that an inflammable gas could be obtained by heating coal in an airtight vessel[6]. Among several recorded examples of the use of coal gas for lighting in Europe in the latter part of the eighteenth century was the earliest use in the UK. In 1792 William Murdock, a Scottish engineer working with Boulton and Watt steam engines for pumping water out of the Cornish tin mines, used coal gas to light a room in his home in Redruth[6]. Murdock subsequently worked on the development of gas lighting with Boulton and Watt's Birmingham foundry in the first decade of the nineteenth century. The gas industry developed steadily in the UK until by the end of the century the amount of coal used in gas production was just over 13 million tonnes, more than for the whole of the rest of Europe[6]. The production process was sometimes known as pyrolysis or carbonization.

The early coal gases were a mixture of various different gases, depending on the nature of the process and the quality of coal. As both gaseous and liquid products were produced from the volatile components in the coal the process was originally known as destructive distillation. The volatile components represented about one third of the original coal and the remaining two thirds formed a solid residue consisting largely of carbon and known as coke. Modern developments in coal gasification were stimulated by the problems of natural gas resource depletion in the United States during the 1960s and by the worldwide availability of enormous coal resources. In these gasification processes coal reacts with steam, forming carbon monoxide, hydrogen and other products. The mixture of carbon monoxide and hydrogen is called synthesis gas. The reaction can only occur when a heat source is applied and some coal is simultaneously reacted with oxygen, releasing heat. The synthesis gas may be used as a fuel or purified and processed to obtain products including hydrogen, methane (SNG), ammonia or liquid fuels. Manufactured gas is classified into three grades with a range of calorific values. In the United States these are known as High, Medium or Low Btu gas[16].

(i) **Low Btu gas.** This is also known as producer gas and is made in a continuous process by passing air or an air-steam mixture over a bed of hot coal or coke. Its calorific value lies in a range from 3 to 6 MJ m^{-3} because of the considerable nitrogen content. The combustible components consist of carbon monoxide and hydrogen. It is uneconomical to transport and can only be used locally for power generation or industrial uses.

(ii) **Medium Btu gas.** The heat generating reaction uses oxygen rather than air and calorific values range from 10 to 22 MJ m^{-3}. At the lower end of the range the gas consists of carbon monoxide, a small amount of carbon dioxide and hydrogen. At the higher end the calorific value is raised by the addition of methane or other hydrocarbons. This gas can be particularly valuable in some manufacturing industries as it burns rapidly and can produce a flame temperature higher than natural gas. It cannot be put directly into the natural gas pipeline distribution system because of its high carbon monoxide content.

(iii) **High Btu gas.** This gas is synonymous with substitute natural gas (SNG) and has a calorific value of about 37 MJ m^{-3}. It is produced by the methanation reaction from carbon monoxide and hydrogen and consists mainly of methane with very small amounts of carbon monoxide, carbon dioxide and nitrogen. It is interchangeable with natural gas and can be added to the natural gas pipeline distribution system.

Many coal gasification proceseses are under development, reflecting the different types and grades of coal as well as the complex chemical conversion processes. Of these the Lurgi process is probably the best known, forming a stage in the South African Sasol Plants described in the previous chapter. Several Lurgi plants are in operation in Eastern European countries[17]. Other commercially available processes for synthesis gas include the Winkler fluidized bed for fertilizer production and the Koppers-Totzek gasification process for ammonia[17]. Methane is also found in coal seams and is collected at the surface by drilling into the gas-bearing strata as the coalface progresses. In some cases it has been available in sufficient quantities for use outside the mine. In the UK methane is also being collected from some municipal wastes at landfill sites for local industrial uses. A survey carried out by the National Coal Board showed that there was a potential annual market for coal-derived fuel gas equivalent to about four million tonnes of coal, mostly from a large number of small scale users[18]. The production of SNG could start to become important again in the UK perhaps during the first or second decade of the next century as North Sea Gas production starts to decline.

Underground Gasification

Underground gasification produces gas directly from the coal seam. The principles were first suggested by Sir William Siemens in 1868 and various unsuccessful attempts have been made since then to achieve commercial production, including extensive work during the 1930s in the USSR[17,19]. Holes are drilled into a coal seam and a controlled fire is started, fed with a suitable combination of oxygen, air or steam, so that a producer gas is formed in the combustion zone. Among problems which have never been satisfactorily overcome are the control of temperature and the extent of combustion. The output gas has variable characteristics and because of its relatively low heating value would be uneconomical to transport and could only be used for local applications. Most of the carbon in the coal is either consumed during combustion or left underground as coke. The WOCOL report[11] pointed out in 1980 that very little progress had been made since the 1950s and that prospects for successful commercial applications were poor. It was not anticipated that underground coal gasification could have any significant impact in coal use before the end of the century. Foley[19] also expressed considerable reservations a few years earlier when he commented

> ...the future for underground gasification of coal is therefore not promising. One, possibly decisive, objection is that it requires a uniform, predictable, regularly sloping coal seam if it is to have any chance of success... (the) kind of seam which can most readily be extracted by machine.

This would put any potential underground gasification process in direct competition with the most easily mined and profitable operations in conventional coal mining.

HYDROGEN

Hydrogen is the lightest chemical element with an atomic weight of 1.0080. It is not an energy source or primary fuel. It has to be produced from other primary substances and the production process uses more energy than the resulting hydrogen can provide. Although it was first recognized in 1766 by Cavendish who called the new substance 'inflammable air', it was not until the twentieth century that ideas for its use as an alternative fuel or energy carrier became widespread[20]. During the 1950s and 1960s a number of workers considered the problem of finding alternative fuels in transportation systems. None of the new energy sources such as solar energy, wind or wave power, could be used directly. However, on the 3rd February 1970 Bockris and Triner[20] suggested that hydrogen would be the best fuel and developed the idea to include other energy applications, calling the concept 'A Hydrogen Economy'. The main features of the hydrogen economy are shown in Figure 5.3.

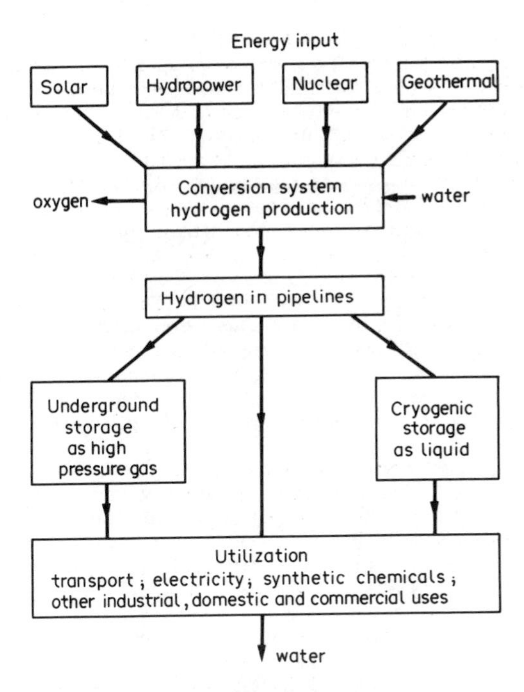

Figure 5.3

Hydrogen forms the intermediary link between the primary energy sources and the energy consuming sectors. It is independent of the primary energy sources used for its production. Even if these need to change the intermediary energy systems of transmission, storage and conversion can continue unaltered. The hydrogen is produced in large plants where the new primary energy sources are available and located away from the point of use. It is then transported either in pipelines or tankers to the end user or put into storage. The hydrogen economy system is completely cyclic. Water from lakes, rivers or oceans is converted into hydrogen and oxygen. Its combustion product is water vapour which is returned to the biosphere and it is the least polluting of all the synthetic fuels. The energy densities (higher heating values) of hydrogen are compared with gasoline (petrol) and methane in Table 5.8. For a given amount of energy hydrogen is about one third of the weight but over three times the volume of the hydrocarbon fuels.

Table 5.8

	Hydrogen	Gasoline	Methane
Gravimetric ($kJkg^{-1}$)	14.2×10^4	4.72×10^4	5.55×10^4
Volumetric (kJm^{-3})	10.1×10^6	34.4×10^6	31.9×10^6

Data from reference 21.

Production

Methods for hydrogen production are mainly based on hydrocarbons at present. The other processes use coal, electricity or high temperature thermochemical decomposition. These methods are reviewed briefly below[22].

(i) **Hydrocarbons.** Natural gas and light oil fractions can be reacted with steam over a catalyst at around 800°C to produce hydrogen, carbon monoxide and carbon dioxide. In the partial oxidation process the heavier oil fractions such as fuel and residual oils can be reacted with oxygen at high temperatures to produce hydrogen and carbon monoxide.

(ii) **Coal.** The main coal gasification routes were described in a previous section. Because a considerable proportion of the original coal is burnt to provide heat for the processes, proposals have been made to substitute nuclear reactors for coal as the heat source. This would save some 50% of the feed coal and reduce the pollution from CO_2, SO_2 and NO_x. Overall cost reductions in hydrogen production are also anticipated.

(iii) **Electricity.** In electrolysis an electric current is passed through water by means of inert electrodes. The electrolyte is usually aqueous KOH. Electrolysis is a well-established commercial technology, but is only used at sites where electricity is inexpensive, for example near hydroelectric power stations. The process is not competitive with hydrogen produced from fossil fuels, but could become a valuable by-product of nuclear electricity at times of minimum electricity demand.

(iv) **Thermochemical Decomposition.** The attraction of directly dissociating water into hydrogen and oxygen through the direct use of heat is that the process is inherently more efficient than electrolysis because the primary energy source does not have to be converted into electricity. Theoretically at a temperature of 3000 K and a pressure of one atmosphere 20% of hydrogen would exist in an equilibrium mixture. The major problems are the properties of materials which would be thermally stable at such high temperatures and the maintenance of the separated hydrogen upon cooling[20]. One method which shows considerable promise is the use of a solar furnace, such as the one at Odeillo, where it has been shown that a mixture of hydrogen and oxygen can be delivered[23].

Storage

Hydrogen has excellent storage properties and can be stored in a liquid or gaseous state or in the structure of solids. Storage as a liquid requires cryogenic storage vessels, but has the advantage that, compared with gas, large quantities can be stored in small volumes. The very large vessels with a capacity of 3200 m^3 used for the United States National Aeronautics and Space Administration's programmes can store liquid hydrogen for several years because of the low evaporation losses[24]. The low density of hydrogen at ambient temperatures makes storage as a gas a major disadvantage, since high pressures and large storage volumes are needed. Underground storage is basically suitable for hydrogen as it has been successfully used for both natural gas and town gas. Cryoadsorption stores hydrogen at medium pressures, up to 40 bar and temperatures around -220°C by physical adsorption on suitable adsorbents, but this process has not yet reached commercial development. Hydrogen can also be stored in chemical combination with metals as metal hydrides. In metal hydrides hydrogen atoms are incorporated in the crystal lattice of metals or alloys, and each lattice can absorb up to three hydrogen atoms[24]. The storage volume is similar to liquid hydrogen storage, but the relatively greater mass of the hydride is a disadvantage. For the same quantity of hydrogen, storage in high pressure gas vessels or hydride storage is about thirty times more expensive than storage in liquid hydrogen containers[24].

Photobiological and Photochemical Hydrogen Production.

The economic photobiological route to hydrogen production is a mid to long term goal and a very considerable amount of basic and applied research is still needed. None of the currently identified photobiological hydrogen producing systems have yet reached the development point at which pilot plant operations would be of benefit[25]. Two examples illustrate the wide range of this work. Among many research centres studying the photobiological production of hydrogen from water is King's College, London, where chloroplast membranes, hydrogenase enzymes and other biological or synthetic catalysts have been investigated[26]. In the United States the University of Miami have been investigating the use of marine and non-marine species of photosynthetic bacteria for hydrogen production[27], a route that was first reported in the 1930s.

The direct conversion of solar energy into stored chemical free energy has attracted research workers for many years[28]. Approximately half the total solar radiation which reaches the earth arrives as visible light and can be used in various photochemical reactions. The other half, which occurs in the infra-red region, cannot make a useful contribution as its energy concentration is too low. The maximum overall efficiency of any photochemical energy conversion is limited to about 30%, however, as a proportion of the higher energy photons of shorter wavelengths have some energy degraded as heat during the reaction. The majority of photochemical reactions are exothermic - giving out energy - and are not suitable for converting solar radiation into stored chemical energy. The known endothermic - energy storing - reactions which occur with visible light are, in theory, capable of producing valuable chemical fuels, but a major problem has been that most of these endothermic reactions reverse too quickly to store the energy of the absorbed light. Other problems include undesirable side reactions and the high cost of the relatively scarce original material. This is relatively unimportant as the original material would be regenerated when the reaction is reversed and the stored energy is released.
One particular process which has attracted attention for many years is the possible combination of carbon dioxide and water to produce various hydro-carbons such as methane. Another possibility is the photosensitised decomposition of water to hydrogen and oxygen. This has been achieved, but with very low efficiencies. In a survey of photochemical research in the early 1980s Porter[29] stated that although a complete photochemical solar fuel generator had not yet been developed, even on a laboratory scale, when such a reactor had been developed then scaling up to cover many square kilometres should not involve many new problems.

Transport Applications

One of the major problems facing the world in the next century is finding an alternative fuel for transport. A considerable number of hydrogen powered vehicles have been developed in the past decade, particularly in the United States. A major difficulty is hydrogen storage in the vehicle. Storage as a compressed gas could be difficult because of the high pressures involved and safety considerations. At pressures around 200 bar the amount of hydrogen stored would represent only 1% of the weight of the storage tank, but its volume would be up to twenty times greater than that of an equivalent gasoline tank. For a given amount of energy stored in a vehicle, hydride storage needs about the same volume as liquid hydrogen storage but can be some 10 to 20 times heavier[24]. It appears that initially hydride vehicles should be considered as alternatives to electric vehicles and used in short range urban transport. However the extra weight would be a disadvantage, causing an increase in fuel consumption which corresponds to the extra energy which would be needed to liquify hydrogen. One of the problems in introducing an alternative road transport fuel is the development of an adequate infrastructure. A network of hydrogen filling stations would be necessary, together with the supply system for the stations. The safe handling of hydrogen in industry is an established technique, but public acceptance of the widespread use of hydrogen could be more difficult to achieve as it is well known to be highly inflamable.

Although a number of studies have been made on the use of liquid hydrogen as a fuel in air transport, only the supersonic sector would appear to be promising. Wilkinson[30] commented that because the aircraft industry still believed (in 1983) that conventional fuel would be available for transport purposes until well into the next century, studies of hydrogen powered airplanes were essentially medium to long term work.

The Hydrogen Economy

An increasing number of professional engineers and scientists believe that the hydrogen energy approach provides the best link between new primary energy sources and the user. Hydrogen could be used in almost every engineering application where fossil fuels are used including direct electricity generation through fuel cells (described in Chapter 10). The main advantages are environmental, as hydrogen does not pollute, and many studies suggest that it would be an economical path also. On the other hand upgrading the enormous reserves of tar sands, oil shale and coal to synthetic crude oils would require large quantities of hydrogen and oxygen, but these could be obtained from electrolysis and from gasifying waste carbonaceous residues[31]. The very serious environmental consequences of such an approach would need careful assessment. This approach would also postpone the full introduction of the hydrogen economy until the latter part of the next century at the earliest.

Summary

Natural gas can play an increasingly important role as a major world energy resource until at least the end of the first decade of the next century. Production could continue to increase at the historic 2 to $2\frac{1}{2}$% per annum growth rates experienced from 1973 to 1983 for at least twenty years. Ultimately recoverable resources are now considered to be broadly comparable with oil, but with the important difference that less than 15% of these ultimately recoverable natural gas resources had been consumed by 1984. The widespread distribution and storage networks which have been established in many countries make the development of synthetic natural gas (SNG) very important.

For the medium to long term view, hydrogen can be considered to be the ideal fuel. However a transition to a possible hydrogen economy is unlikely to take place over the next few decades because the traditional fossil fuels will still be widely available. Many new and renewable routes to hydrogen production are being investigated but it is too early to say if these can ever become commercially viable.

References

1. Tiratsoo, E.N., Natural Gas, Second Edition, Scientific Press Ltd, London 1972.

2. Natural Gas, Briefing Paper, British Petroleum Company, September 1981.

3. Heron, D., Paper 93, 8th Comm. Min. and Met. Congress, Melbourne, 1964 quoted by Tiratsoo[1].

4. Cook, Earl, Man, Energy, Society, W.H. Freeman and Company, San Francisco, 1976.

5. Woytinsky, W.S., and Woytinsky, E.S., World Population and Production. Trends and Outlook, 917-931, New York, 1953.

6. Williams, T.I., A History of the British Gas Industry, Oxford University Press, 1981.

7. Hubbert, M. King, Energy Resources, in Resources and Man, National Academy of Sciences, W.H. Freeman and Company, San Francisco, 1969.

8. B.P. Statistical Review of World Energy 1982, The British Petroleum Company plc, London, May 1983.

9. B.P. Statistical Reviews of the World Oil Industry, 1975-1980 inclusive, The British Petroleum Company Limited, London, 1976-81.

10. Stern, J.P., Soviet Natural Gas Development to 1990, Lexington Books, Massachusetts, 1980.

11. Rahmer, B.A., Reserves and Resources, Petroleum Economist, 329-334, September 1983.

12. Halbouty, M.T. Reserves of Natural Gas Outside the Communist Block Countries, World Petroleum Congress, London, 1983.

13. Worldwide Report, 80-81, Oil and Gas Journal, 26 December 1983.

14. Ion, D.C. Availability of World Energy Resources, Graham and Trotman Ltd, London, 1975.

15. Oil and Gas in 1982, Shell Briefing Service No. 2., Shell International Petroleum Company Ltd., London, 1983.

16. Bodle, W.W. and Huebler, J., Coal Gasification, in Coal Handbook, Marcel Dekker Inc., New York and Basel, 1981.

17. Coal-Bridge to the Future, Report of the World Coal Study (WOCOL), Ballinger Publishing Co., Cambridge, Massachusetts, 1980.

18. Gas from Coal, National Coal Board, London, 1983.

19. Foley, Gerald, The Energy Question, Penguin Books, 1976.

20. Bockris, J. O'M., Energy Options, Taylor and Francis Ltd., London, 1980.

21. Veziroglu, T.N., Hydrogen Energy System Concept and Engineering Applications, Int. Conf. Future Energy Concepts, Institution of Electrical Engineers, London 1979.

22. Steering Committee on the Production of Hydrogen from Water, European Federation of Chemical Engineering, 1977.

23. Lede, J., Lapicque, F. and Villermaux. J., Production of hydrogen by direct thermal decomposition of water, Int. J. Hydrogen Energy, 8,9,675-679, 1983.

24. Peschka, W., Energy storage with hydrogen, in Hydrogen: Energy Vector of the Future, Graham and Trotman Ltd., London, 1983.

25. Weaver, P.F. Lien, S. and Seibert, M., Photobiological production of hydrogen, Solar Energy, 24, 3-45, 1980.

26. Gisby, P.E. Rao, K.K. and Hall, D.O., Hydrogen Production from water, using chloroplasts, enzymes and synthetic catlysts: a photobiological process, Solar World Forum, Vol. 3, 2242-2247, Pergamon Press, Oxford, 1982.

27. Mitsui, A. et. al., Photosynthetic Bacteria as Alternative Energy Sources: Overview on Hydrogen Production Research, in Alternative Energy Sources II, 3483-3510, Hemisphere Publishing Corporation and McGraw-Hill 1981.

28. McVeigh, J.C., Sun Power, Second Edition, Pergamon Press, 1983.

29. Porter, G., Prospects for a biological synthesis of 'biomass', Paper V/KI, Proc. Conf. Energy from Biomass, Applied Science Publishers, 1981.

30. Wilkinson, K.G., An airline view of LH2 as a fuel for commercial aircraft, Int. J. Hydrogen Energy, 8, 10, 793-6, 1983.

31. Shannon, R.H. and Richardson, R.D., Resource and Energy Management of Synfuels, Int. J. Hydrogen Energy, 8, 10, 783-792, 1983.

Further Reading

Dr. E.N. Tiratsoo's Natural Gas, published by the Scientific Press, London, second edition 1972, is probably the classic scientific text, covering all the developments up to the early 1970s. For the UK Dr. Trevor Williams has written A History of the British Gas Industry, Oxford University Press, 1981. This starts with the discoveries of the eighteenth century and ends with North Sea Gas firmly established. The Public Relations Department of the British Gas Corporation, 152 Grosvenor Road, London SW1V 3JL, can provide a wide variety of useful literature on current developments. Coal Handbook, edited by Robert Meyers, Marcel Dekker Inc., New York and Basel 1981 has an excellent section on gasification.

Energy Options, by J. O'M. Bockris, Taylor and Francis Ltd., London, 1980 gives a comprehensive study of the world's future energy resources and develops the case for a solar-hydrogen economy. A very detailed survey of photobiological hydrogen production containing over four hundred references was given in a review paper Photobiological Production of Hydrogen, by P.F. Weaver, S. Lien and M. Seibert in Solar Energy, 24, 3-45, 1980. The International Journal of Hydrogen Energy, published by Pergamon Press, covers all aspects of hydrogen energy, including economic and environmental issues. Volume 8, number 9, 1983, contains a major solar-hydrogen book list and bibliography.

Exercises

1. The table shows world cumulative production of natural gas using the assumption that actual production is 20% greater than consumption. Show that these figures are consistent with the logistic curve

$$Q = \frac{9600}{1 + e^{2.332-0.074t}}$$

where t = 0 in 1974 and hence show that annual production will next drop to the level of 1974 in the year 2047 if these trends continue.

	Cumulative Production Tcf
1974	850
1976	955
1978	1067
1980	1189
1982	1313

2. A comment on the development of natural gas in the Netherlands stated[6]

> It is now generally conceded that the Dutch sold too much gas too quickly and too cheaply, thereby encouraging its use in areas where lower premium or even non-premium fuels would suffice...

   What is meant by 'too much', 'too quickly' and 'too cheaply'? What applications for natural gas could be replaced by less valuable fuels today?

3. In 1979 the rate of natural gas flaring in the UK sector of the North Sea was 18 million cubic metres a day. By the end of 1983 this had fallen to 10 million cubic metres a day. The Minister of State for Energy commented that these figures demonstrated the real national importance of this saving.
   (a) How much gas was being flared in 1979 and 1983 in million tonnes of oil equivalent?
   (b) What percentage of natural gas production does this represent in these years?
   (c) What is the present position regarding the flaring of gas in the North Sea?

4. Examine the development of the UK natural gas transmission system. What are the long term advantages of having such a bulk energy transport system?

5. The Morecambe Gas Field is the first major gas field to be developed in the Irish Sea. How will its resources compare with the earlier fields in the Indefatigable and Leman Fields in the North Sea?

6. What risks are associated with the storage and transportation of LNG? How has the accident rate compared with the transportation of oil at sea?

7. What are the main advantages and disadvantages of hydrogen as a fuel today? How could these alter with suitable development of production and other facilities in thirty years time?

8. Examine the trends in the cost of large scale photovoltaic systems and in the costs of electrolysis. If the cost of conventional transport fuels were to rise by 3% per annum in real terms when would a photovoltaic-hydrogen economy be viable?

9. Examine the present position in the production of hydrogen by marine and non-marine species of photosynthetic bacteria. How near is this to commercial application?

Answer

3. (a) 5.913; 3.285; (b) 17.5%; 10.1%.

6. HYDROPOWER AND BIOMASS

Introduction

The various traditional designs of watermill used until the nineteenth century could only lead to a technical dead end[1]. None of them were capable of using a head of water much greater than their own diameter. Further progress followed with the development of the water turbine which was subsequently linked to an electric generator. Although credit for the world's first hydro-electric plant is often attributed to the United States plant which started in the autumn of 1882 at Appleton, Wisconsin, two plants were already operating in the UK at that time[2]. The earliest was Sir William Armstrong's small hydro-electric plant, rated at just under 5kW, which was constructed in 1880 to light his picture gallery at Cragside, Northumberland some 1.5 km away. The first public supply of electricity was reported from Surrey in 1881, when electric current generated from the waters of the River Wey was used to light the streets of Godalming. The cables had to be laid in the gutters as there was no legal authority to dig up the streets.

The world's first large hydro-electric plant was built in 1895 at the Niagara Falls in the United States, with two turbines each rated at 4100 kW[3]. The subsequent development of alternating current by George Westinghouse in 1901 allowed electric power to be transmitted over long distances[4]. By 1903 Canada had a 9.3 MW plant, also at Niagara Falls, and the era of modern hydropower had commenced. The first reliable survey of water turbines manufactured and installed throughout the world in the late 1920s[5] suggested that about 40% of the world's electricity was generated by hydropower, with the United States and Canada having a combined operating potential capacity of over 13 000 MW and five other countries, France, Japan, Norway, Sweden and Switzerland, with operating potential capacities greater than 1000 MW. A few of the earlier hydropower plants, known as run-of-the-river plants, could not generate any power when the river was low during the dry season, but by the 1930s the use of large dams had been established in the United States. The creation of the Tennessee Valley Authority in 1933 with their comprehensive approach to the planning and development of river basins set a pattern which has been widely followed in other countries[4]. Since then there has been a steady growth in hydropower throughout the world although the percentage share of hydropower in meeting world electricity demand had fallen to about 25% by the early 1980s. In 1982 North America and Europe accounted for nearly 60% of the total world hydropower production, while among the developing countries Brazil, Ghana, Mozambique, Zaire and Zambia obtained over 85% of their electricity from hydropower[6,7].

The potential for development of hydropower over the next forty years is so great that it could provide an output equivalent to the total electricity generated in the world from all sources in the early 1980s. Much of this potential will be established in the developing countries, some of whom could increase their present use of hydropower by a factor of ten or more.

The Basic Hydropower Plant

The basic principles of hydro-electric power generation are shown in Figure 6.1. Water at a high level, often stored behind a dam, falls through a head z. Its gravitational potential energy is converted to kinetic energy and the flowing water drives a water turbine. The rotating turbine shaft drives the electric generator to produce electricity.

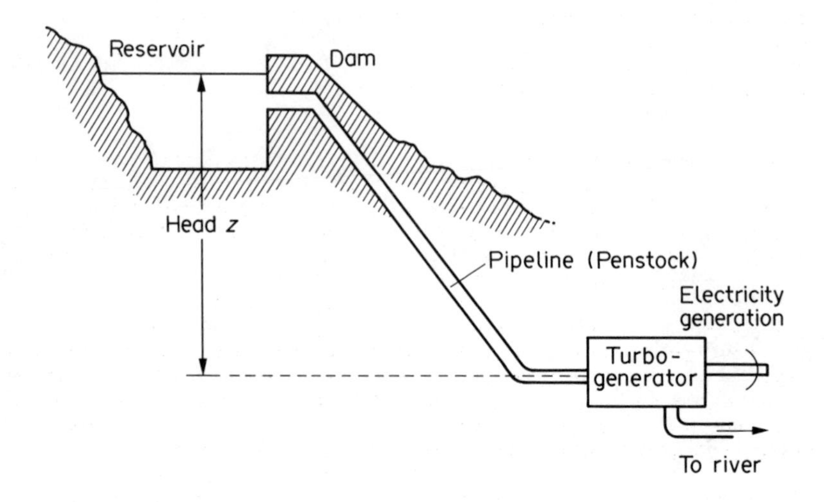

Figure 6.1

Example 6.1 In a hydropower plant the head of water in the dam is 250m. Assuming that the gravitational potential energy can be converted to kinetic energy at the turbine inlet without loss, calculate the maximum velocity of the water.

Solution: Gravitational potential energy = Mgz = Kinetic energy

$$Mgz = \frac{Mv^2}{2}$$

$$9.81 \times 250 = \frac{v^2}{2}$$

$$v = 70 \text{ ms}^{-1}$$

Example 6.2 If the flow in Example 6.1 is 60 m^3 per second and the combined conversion efficiency of the turbine and electric generator is 89%, calculate the output.

Solution: Mass flow = 60 000 kg s^{-1}

Gravitational potential energy = 9.81 x 250 J kg^{-1}

= 2452.5 J kg^{-1}

Output = 0.89 x 60 000 x 2452.5 watts (J s^{-1})

= 131 MW

In real plants there are additional hydraulic losses due to friction in the penstock and at its inlet and outlet. Overall efficiences of hydropower plants are typically in the range from 75 to 85%.

Types of Turbine

Turbines can be classified according to the direction of the water flow through the blades e.g. radial, axial or combined-flow turbines, or as reaction, impulse or mixed-flow turbines. In reaction turbines there is a change of pressure across the turbine rotor, while impulse turbines use a high velocity jet impinging on hemispherical buckets to cause rotation. There are three basic types of turbine broadly related to low, medium or high heads.

Propeller or axial flow turbines are used for low heads in the range from 3 to 30 metres. They can have relatively inexpensive fixed blades, which have a high conversion efficiency at the rated design conditions but a poorer part-load efficiency, typically 50%, at one third of full rated output. Alternatively the more expensive Kaplan turbine has variable pitch blades which can be altered to give much better part-load efficiency, perhaps 90% at one third of full rated output.

The Francis turbine is a mixed-flow radial turbine and is used for medium heads in the range from 5 to 400 metres. It has broadly similar performance characteristics to the fixed-blade propeller type.

The best known impulse turbine is the Pelton wheel. Each bucket on the wheel has a centrally placed divider to deflect half the flow to each side of the wheel. It is normally used for heads greater than 50m and has good performance characteristics over the whole range, very similar to the Kaplan turbine, reaching 60% efficiency at one-tenth of full rated output.

Hydropower Potential

The earth's energy flow diagram, Figure 1.9 on page 17, shows that just over 4 x 10^{16}W flows in the hydrological cycle of evaporation, rain, other precipitation and storage in water and ice. A very small proportion of this hydrological energy flow, probably between 0.01 and 0.015%, is considered to be theorectically available for conversion into hydropower[6.8]. This theoretical world hydropower potential is calculated as the total energy potential of river discharges relative to a datum of sea level or the base level of erosion for closed basins and is widely quoted as 44.28 x 10^{12} kWh per annum[6.8]. However, this figure does not seem to include the 3.94 x 10^{12} kWh for the USSR, which is separately listed by the 1980 World Energy Conference[6] and there is also some doubt as to whether or not the 6×10^{12} kWh estimated for the People's Republic of China has been included[8]. A better assessment is the 'technically usable hydropower potential', which allows for the unavailability of certain river reaches, mainly those near estuaries. This is less than half the theoretical value. The 'economic potential' includes all hydropower resources which are regarded as economic compared with alternative sources of electric power at the time of the assessment. These can be classified into three categories: operating, under construction and planned, as shown in Table 6.1, derived from 1980 data[6]. The economic operating hydropower potential of 372.1 GW represented just under 16% of the technically usable potential, and supplied about 23% of the world's electricity.

Table 6.1

Hydropower Potential (GW)

Region	**Technically Usable**	**Economic**		
		Operating	**Under construction**	**Planned**
Asia (1)	609.6	53.1	9.1	42.0
Latin America	431.5	34.1	40.5	92.4
Africa	358.4	17.2	5.4	22.9
USA and Canada	356.2	128.9	34.6	39.0
USSR	250.0	30.3	21.8	19.4 (3)
China (2)	216.9	5.7	5.9	Unknown
Europe	163.2	96.1	10.7	22.5
Rest of world	44.5	6.7	2.3	3.6
Total	2430.3	372.1	130.3	241.8

(1) Figures from Asia probably do not include data from the People's Republic of China[8].
(2) Figures may not include all small hydropower plant.
(3) Estimated

The world operating potential of some 372.1 GW could, in theory, have provided $372.1 \times 10^9 \times 365 \times 24$ watt-hours or 3.26×10^{12} kWh in that year (1979).
The actual energy generated in 1979 was 1.65×10^{12} kWh[7]. This represents 50.6% of the potential, a typical figure for most hydro-electric plant. Not only are there seasonal fluctuations in water availability, but the demand for electricity fluctuates and plants need to close for maintenance. In the United States and Canada the figure of (actual energy generated) divided by (theoretically available potential) was 47.7% in 1979. This ratio is known as the load factor. As the electrical power from a hydro-electric plant can be used directly without the conversion losses and wasted heat associated with conventional fossil fuel power plant, the primary energy equivalent of hydro-electricity is usually taken as about three times its actual output (see pages 9 and 10). A common conversion is that four thousand kWh of 'electricity generated' is considered to have the primary energy equivalent of one tonne of oil[7]. Hydropower represented 6% of the equivalent primary energy consumption in the world in 1979 and was estimated to produce some 23% of the world's electricty.

There has been a steady growth in hydropower for many years at about 3.5% per annum, representing a doubling period every twenty years. This figure was used by the

1980 World Energy Conference to estimate that hydropower could be quadrupled by 2020. This is examined in Figure 6.2,

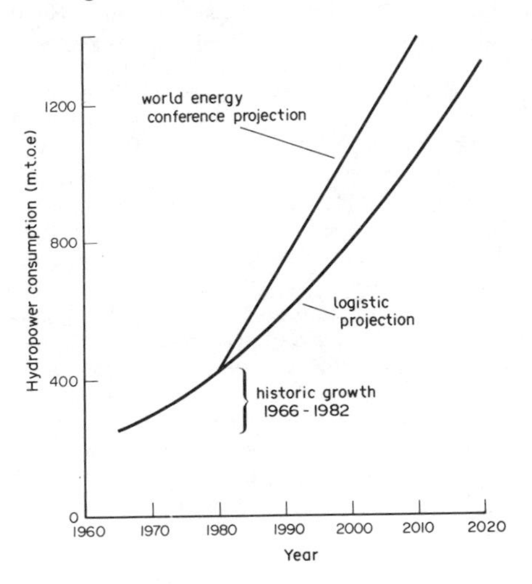

Figure 6.2

which shows the historic growth in hydropower consumption, in million tonnes of oil equivalent, the World Energy Conference projection and the use of a projection based on the logistic equation

$$\frac{Q}{Q_f} = \frac{1}{1 + 9.0971e^{-0.04082t}}$$

which was derived from the data[7,9] given in Table 6.2. taking t = 0 in 1966:

Table 6.2

World Hydropower Consumption (million tonnes of oil equivalent)

Year	Consumption
1966	260.0
1968	275.2
1970	306.1
1972	326.6
1974	348.6
1976	362.7
1978	402.6
1980	420.2
1982	446.0

Q_f was calculated to be 2625 m.t.o.e. by assuming that the technically usable 2430.3 GW would be available for just under 50% of the year. The logistic curve projection is probably more realistic for reasons already outlined for limited growth in Chapter 1.

Pumped Storage

Pumped storage systems are used at times of peak demand for electricity. The water can be pumped to an upper storage reservoir usually at night when the demand is low, and can then be allowed to flow down through the turbines, generating electricity when it is required. The largest pumped storage system in Europe has just been completed in the UK at Dinorwig, near Llanberis in North Wales. During its construction 3 million tonnes of rock were excavated from the heart of the mountain between two reservoirs and 16 km of shafts and tunnels were created[10]. The upper reservoir is 568 m above the underground power station and the horizontal distance between the upper and lower reservoirs is 3200 m. There are six turbo generator units, each rated at a nominal 300 MW. It can generate at full output for about five hours. The overall efficiency of any pumped storage system is less than the 'once through' conventional plant, as the pumping efficiency during the return flow to the upper reservoir must be included. This pumping efficiency, typically about 90%, reduces the overall efficiency to about 70 to 75%. However, the economics are quite different. Pumping to the upper reservoir only occurs when the electricity tariffs are low. Electricity is supplied to meet peak demands when tariffs are usually at their highest. Further, the use of a pumped storage system reduces the need for additional conventional plant which would only be needed for very short periods each year. The pumped storage plant at Dinorwig can be generating electricity within ten seconds of requirement.

Small Scale Hydropower

One of the needs in many parts of the world is for electrical power in remote regions far from a conventional transmission system. Small scale hydropower is again becoming considered for an increasing number of these applications. The early history, up to the 1930s, was largely dominated by small plants, less than 1 MW in capacity, but then the economies of scale began to favour large-scale development. Until fairly recently it was necessary to match the turbine design very carefully to the particular site. This resulted in an expensive special 'one-off' hydropower generator. The smaller the application the greater the installed cost per kilowatt of capacity. The need for these specially designed systems has been largely overcome by the use of standardized turbines and associated equipment, with the acceptance of some loss in overall plant efficiency and performance.

One of the major factors which could favourably influence the economics of small-scale hydropower is the development of microprocessor-based electronic load governors. These can overcome problems of instability in matching waterflow to a variable demand and can also reduce costs as expensive mechanical controls are no longer necessary. By the 1980s the country with the greatest experience in small-scale hydropower development was the People's Republic of China, where nearly 100,000 plants have been constructed in the past twenty years[8]. Most of their recent plants have a rated output of some 300 kW and much of their projected increase in hydropower over the next twenty years will also be small-scale.

Economic, Social and Environmental Issues

The costs and benefits of hydropower plant are usually evaluated by an economic comparison with conventional thermal or nuclear power stations. The main factors which must be considered in addition to increases in construction costs, are changes in the cost of fossil fuels and in environmental protection regulations. Although there has been a steady growth in power station construction costs in all countries over the past two decades, thermal and nuclear power station costs have risen at a greater rate than those of hydropower plants. There are two reasons for this. The technology and management of the construction of hydropower plants has improved relative to conventional power station construction and new environmental protection and safety regulations have adversely affected the cost of nuclear and coal-fired power stations[8]. These new regulations have resulted in greatly increased expenditure for the control of air and water pollution with coal-fired stations, and for radiation monitoring and control together with improved safety standards in nuclear installations. Some of the adverse effects of hydropower schemes, such as the essential reinforcement of river banks or compensation for moving and resettling whole communities from flooded land, have always been included in the overall construction costs. It is considered very likely[8] that the rate of increase in the cost of nuclear and coal-fired power stations over the next fifteen years will continue to be greater than the corresponding rate of increase for hydropower plant. This will increase the competitiveness of new hydropower projects and extend the range of economically usable potential. Much could also be done to increase hydropower capacity without building new plant. Dams with existing hydropower schemes can be raised to provide both additional storage capacity and a potentially increased output. Turbine generators can be added to some existing storage reservoirs to create new generating capacity.

Several analysts have drawn attention to the role of hydropower in the problems facing the oil importing developing countries. For example in Brazil in 1983 the hydropower capacity was 35,495 MW, representing some 83% of total electricity generation. This is planned to increase to 41,786 MW of hydropower by 1986, but their electricity supply authority is in a very difficult financial position resulting from the large external loans which were necessary to build the hydropower plants. The need to reduce the 40% share of oil in the total Brazilian energy consumption has resulted in a number of incentives to industry, particularly for the use of off-peak electrical power and the use of hydropower during the months when the rivers are full and not in the dry season[11]. In Sri Lanka the Mahaweli project is planned to provide at least a further 500 MW of hydropower capacity to an installed capacity in 1984 of 562 MW, increasing the percentage of hydro-electricity to over 80%. The reasons for the accelerated development of the Mahaweli project to a seven year project instead of the thirty years advocated by the United Nations Development Programme are also directly related to the price rises in oil. As recently as 1982 the Ceylon Electricity Board paid 63% more for its fuel than in 1981[12]. In parallel with the need for increased electrical capacity the importance of integrating irrigation schemes into the overall development plan has been recognised by the inclusion of several complementary irrigation water diversion and distribution projects in the Mahaweli project.

Rivers and streams are regarded in the great majority of countries throughout the world as a public resource. Their use in potential hydropower schemes is subject to government control. Hydropower development may be socially acceptable to some sectors of the community and have quite disastrous effects on others. For example the construction of the Aswan High Dam in Egypt resulted in the destruction of the sardine fishing industry in the Eastern Mediterranean, but this was balanced by the development of a new fishing industry on the newly created Lake Nasser[4]. There have been many studies on the adverse impacts on health which can result from the large dams associated with hydropower projects[4] and it would appear that there is still a need for major health education programmes to be associated with these projects, so that diseases such as bilharzia and malaria could be eliminated. Other associated environmental problems include the need for extensive drainage systems on newly irrigated land and the threats to new dams caused by widespread deforestation and soil erosion many kilometres upstream. Some existing aquatic and terrestial ecosystems have been disrupted and there may have been a loss of visual amenities in scenic areas. On the other hand the United Nations Hydropower Panel[8] has also drawn attention to the positive effects of hydropower reservoirs on the environment. The creation of regulating reservoirs has been shown to make a substantial improvement in the water supply for domestic, industrial and agricultural purposes in many cases. The danger of catastrophic floods has often been eliminated. The overall effects of hydropower schemes throughout the world have been beneficial, although there have been some largely unanticipated adverse reactions with the environment. These could either be reduced or eliminated through careful resource planning.

BIOMASS

Introduction

Historically the development of man can be directly traced through biological conversion systems, initially through the provision of food, then food for animals, the materials for housing and energy for cooking and heating. The commencement of industrial activities was followed by the development of agriculture and forestry to their present levels. The renewed emphasis on biological conversion systems arises from the fact that solar energy can be converted directly into a storable fuel and other methods of utilizing solar energy require a separate energy storage system. The carbohydrates can be reduced to very desirable fuels such as alcohol, hydrogen or methane, a process which can also be applied directly to organic waste materials which result from food or wood production.

Biomass can be defined as all types of animal and plant material which can be converted into energy. It includes trees and shrubs, grasses, algae, aquatic plants, agricultural and forest residues, energy crops and all forms of wastes. Estimates of how much of the world's energy demand is met by biomass range from 6 to 13 per cent[13,14]. Among developing countries, biomass is the single most important source of energy especially within the domestic sector although some local industries such as bakeries, brick firing or steam production are also dependent on fuel wood. Nine-tenths of the population in the poverty belts rely on wood as their chief source of fuel, and although in cooler regions some is used for heating, by far the most important energy need is for cooking[15]. Cooking fuel represents approximately 50% of fuel use in many rural areas and up to 90% of energy needs in warmer areas[16].

The quantities of biomass produced throughout the world are very large. The annual net production of organic matter has an energy content of some 3.1×10^{21} J, some ten times the world's annual energy use[13]. In forests alone the biomass productivity is some three times the world's annual energy use[14]. Estimates of the potential biomass resource already standing in the world's forests, 1.8×10^{22} J, are in the same order as the proved world oil and gas reserves[13].

Photosynthesis

Solar energy can be used by all types of plants to synthesise organic compounds from inorganic raw materials. This is the process of photosynthesis. In the process carbon dioxide from the air combines with water in the presence of a chloroplast to form carbohydrates and oxygen. This can be expressed in the following equation:

$$CO_2 + H_2O \xrightarrow[\text{chloroplast}]{\text{sunlight}} C_x(H_2O)_y + O_2$$

A chloroplast contains chlorophyll, the green colouring matter of plants. The carbohydrates may be sugars such as cane or beet, $C_{12}H_{22}O_{11}$, or the more complex starches or cellulose, represented by $(C_6H_{10}O_5)_x$. All plants, animals and bacteria produce usable energy from stored carbon compounds by reversing this reaction. Compared with other methods biological or photosynthetic conversion efficiencies are much lower, but are potentially far less expensive. Photosynthetic efficiency is based on the amount of fixed carbon energy produced by the plant compared with the total incident solar radiation. Plants can only use radiation in the visible part of the solar spectrum between wavelengths of 400-700 nm, known as the photosynthetically active radiation (PAR) region. This represents about 43% of the potentially available total radiation. At the plant some of the PAR is reflected and with other losses due to internal chemical processes the maximum attainable efficiency lies between 5 and 6%.

Under very favourable conditions, conversion efficiencies of between 2 and 5% have been recorded in the field for growth periods of a few weeks, but considerably lower efficiences are achieved over longer periods of growth. Irish grasslands or forests with Sitka spruce are capable of dry matter yields greater than 16 tonnes/ha which represents an efficiency of about 0.7%. The main reasons for these relatively low efficiencies are environmental constraints, nutritional limitations and the incidence of pests and diseases[17]. Typical environmental constraints would include a drought or daily variations in ambient temperature. Nutritional limitations depend on the soil quality which, in turn, relies on the output of fertilizers.

Energy Resources

There are five routes which can be followed to obtain the organic material or biomass which is the starting point for the energy conversion process. The first, and by far the simplest, is to harvest the natural vegetation. There are fertile regions in many parts of the world where the topography or some other reason makes the land unsuitable for agriculture or other valuable activities. With the harvesting of natural vegetaion, no costs are involved in planting or clearing and the land would be given a new use. A major disadvantage of this method is that the yields are at best about half those which could be otained from an energy plantation. The second is through the cultivation of a specific energy crop, grown only for its energy content, or the use of agricultural surpluses, so that the stored chemical energy can be converted into useful energy by combustion or coverted into a storable fuel. A land crop should have as high a conversion efficiency as possible, but it does not have to be digestible by animals or edible by humans. The entire material or biomass of the crop can be used, including the leaves, stalks and roots. By careful genetic selection and intensive cultivation the conversion efficiency should reach 3% under normal conditions. In the third route, trees and other types of lignocellulose material are grown specifically as fuel in energy plantations. Short rotation forestry, described later, is a good example. The fourth uses the wastes from agro-industrial processes or residues from agriculture, animal wastes, straw, and all forms of urban wastes. The fifth route is through algae in the sea or grown in inland ponds.

Conversion of Biomass to Fuels and Other Products

A selection from some of the main conversion processes is illustrated in Figure 6.3 which shows that there are often several different routes to the same end product. Combustion is by far the simplest and best known technique, particularly with forestry residues and industrial and urban wastes. A number of the processes are well known and are ideally suitable for producing fuels. With aerobic fermentation, materials containing starches and simple sugars can be used to produce ethyl alcohol or ethanol. Aerobic fermentation has the added advantage of producing a valuable by-product, the nutrient-rich fertiliser from the digested slurry, when used to treat domestic sewage or animal wastes and produce biogas. In the pyrolysis process the organic material is heated to temperatures between 500 and 900°C at ordinary pressures in the absence of oxygen, producing methanol, which was a byproduct of charcoal in the last century. Methanol was first used as a fuel for high performance racing cars and was subsequently studied as an additive in many laboratories. It is now considered to be an essential part of the future automobile fuel mixture[18].

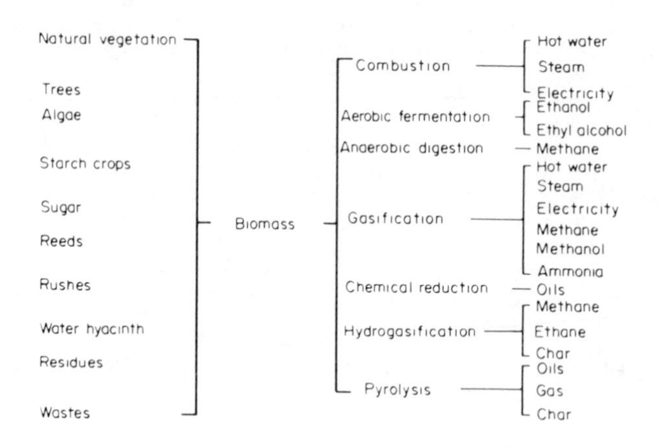

Figure 6.3

Cooking - the major application of Biomass

In a recent United Nations Conference on Energy it was reported that 'more than 90% of wood cut in Africa - five million acres a year - is burnt as fuel'[19]. Deforestation and desertification is widespread and increasing. In the Sahel the southern edge of the Sahara has moved some 100 kilometres in 17 years. The scarcity of wood in some areas has meant that local inhabitants have had to move on or turn to substitutes. The World Bank[14] reported that between a half to one billion people use argicultural or animal wastes to fuel their fires. In India cattle dung represents three-quarters of the Indian domestic fuel consumption robbing the land of valuable nutrients[20]. In parts of Africa crop residues and stubble are uprooted and used for fuel.

In short, there is now a severe shortage of biomass for cooking. According to the Food and Agricultural Organization of the United Nations[19]

> 100 million people are now unable to obtain sufficient wood to meet their needs; a further one billion are affected by shortages

Cooking is a very cultural-specific activity. However the most common means of cooking throughout the developing world is on an open fire. The most basic stove is simply three stones arranged on the ground in a triangle. The pan rests on the stones, between which three or more pieces of wood are placed. Efficiencies are low, between 2 and 10%, although much depends on the rate of burning, the air convection and other factors such as the height of the pan above the fire. To some extent the rate of burning can be controlled by pushing the sticks closer together for a fast burn or by pulling them apart for a slower rate of burning. The number of sticks in a fire has a greater effect on the rate of burning than does the volume of the wood. In some parts of Africa the three stone fire is often expanded into a five stone base for two pots. One attempt to improve on the efficiency of firewood use consists of adding a wind protection[21]. Otherwise there are few cheap and simple modifications to the open fire which can significantly reduce wood consumption.

Alternative fuelwood stoves are permanent or monolith structures. In South-East Asia and the Indian sub-continent the use of woodburning stoves made from clay is widespread. They consist of a hollow rectangular box with an opening underneath for the fire and two or more holes on the top for the cooking pots. Known as 'chulas', there are many different versions but they are often no more efficient than a three stone open fire. Singer[22] reported them to be only 6 to 7% efficient. Apparently the cooking pots absorb only a small percentage of the radiant heat from the fire and cold incoming air tends to flow over the top of the wood, lowering the flame temperature and preventing the ignition of the volatile wood gases. Improvement of the clay chulas consists of adding a flue and front damper which controls the availability of oxygen. More sophisticated designs may employ baffles to redirect hot gases. Imported solid stove designs are said to make a fuel saving of 30 to 40% over an open fire[23]. However, clay stoves tend to perform better when long periods of cooking are needed and when all the pot holes are being utilized. In some parts of the world such as East Africa, clay is not generally used for cookstoves. Furthermore, in desertified areas such as Khartoum, which is served from a long distance, charcoal is the predominant fuel since it is easier to transport.

Figure 6.4 shows a traditional charcoal stove made from a rectangular tin with a hole cut in one side and a wire grate attached to the top on which the fuel is burnt.

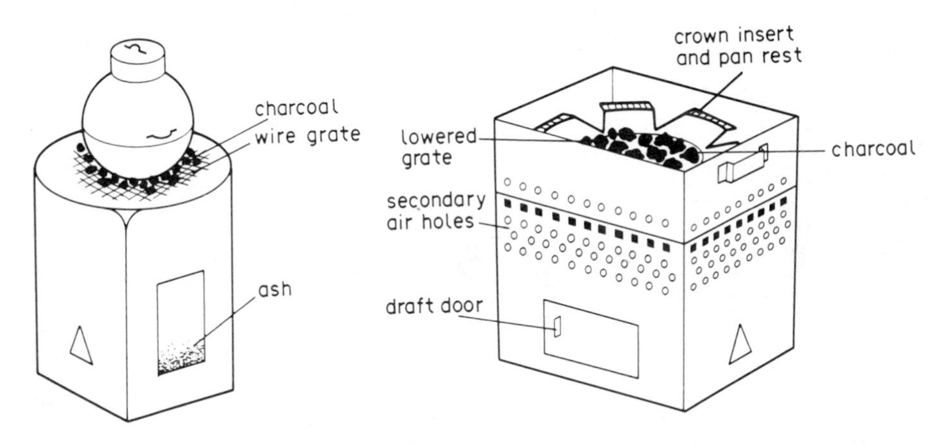

Figure 6.4 Traditional stove

Figure 6.5 Improved stove

The fire is not always confined to the cooking pot, and inevitably heat losses to the surrounding area and from the fire container itself are wasteful. Preliminary tests[24] showed this stove to be about 18% efficient, which is about two to five times more efficient than an open fire. Taking into account wood to charcoal conversion factors in the range of 15 to 30% and the fact that charcoal has approximately twice the calorific of wood, burning charcoal on a metal stove gives approximately twice the heat utilization compared with cooking on an open fire.

Figure 6.5 shows an improvement on this basic stove. Lowering the height of the grate gives a slight 'chimney effect', which gives good aeration to the coals providing fast even combustion. The 'crown insert' limits the fire to the base of the pan, but the hot gases are forced to circulate around the side of the pan. The draft door can be shut down after boiling, reducing the rate of burning. Secondary air holes around the base of the grate preheat incoming air which further helps liberate heat from the coals during the simmering stage of cooking. Using this stove, laboratory tests showed that the same cooking process could be achieved using only 60% of the charcoal required for the unmodified stove, although this only showed as a 4% improvement in stove efficiency. 'Economic heat utilization' is a more preferable measure of performance.

Results from the consumer field trials in Northern Sudan with the improved stove were very favourable[24]. Women commented on the economic use of fuel, the slow, quiet burning fire and the fact that all the heat is concentrated to the pan. Other comments included the ease of lighting the fire, and the fast time to boil. Although the women generally added a few extra pieces of charcoal to complete their cooking, they estimated that fuel saving was in the range of 25 to 33%. However, some difficulties were experienced in quantifying the womens' fuel saving claims. Physical measurements over a long testing period are needed to determine the actual reduction in fuelwood demand that improved stoves could make. Otherwise, perceptions of economy and fuel saving can be complex. For example the ease and speed of lighting the fire and bringing the foodstuff to cooking temperature can initiate a 'halo effect', in that it influences other favourable impressions of a stove with regard to overall fuel consumption. The sociological aspects associated with stove acceptance are at least as important as technical innovation.

Another approach to alleviate the demand for fuelwood is the use of solar energy for cooking. Since many fuel-scarce countries are situated in tropical arid and semi-arid areas, the potential value of solar energy devices has received attention. Several types of solar cookers have evolved but perhaps the most common is that based on the idea of a simple parabolic reflector which focuses the direct solar radiation on a small area. The reflecting material can be spun aluminium[25], aluminized plastic film applied to a parabolic shell as in the case of Von Oppen's sun basket[26] or small glass mirrors[27]. The use of spun aluminium or mirrors is expensive and while aluminiumized plastic film is cheaper, it suffers severely from degradation in sunlight and cannot withstand frequent cleaning under harsh environmental conditions. All direct focusing dishes need to be refocused every 15 to 20 minutes. The hot box cooker is one of the simplest designs to make. Solar radiation enters a well insulated box through glass panes and is absorbed by dull black inside surfaces[28]. The addition of four flat plate collectors augments the solar energy reaching the box and is generally known as a solar oven[29]. Once the oven is put in the sun, few adjustments are required. Two less common solar units are the combination unit and the steam cooker. The former is an attempt to combine the best features of the direct focusing cooker with those of the oven type cooker. A concentrating system is used to focus the sun's rays into an insulated oven[30]. The steam cooker was developed as a means of providing indoor cooking without the need to adjust the collector. A flat plate collector is used to raise steam to an insulated cooking vessel[31].

Simple hot box cookers can only reach fairly low cooking temperatures, perhaps 100°C during summer months, and are best suited to the long slow cooking of rice and pulses such as dried peas. The slow boiling of cereals, rice and potatoes has also been achieved with the solar steam cooker. The solar oven can reach higher temperatures, suitable for stewing, baking and even roasting. Similarly, the direct focusing cookers, can also reach higher temperatures, achieving relatively fast cooking. However, some of the lightweight units have experienced problems of stability in wind. Overall, it is only the direct focusing cooker that has received any extensive field testing. 250 parabolic dish concentrators were introduced to three villages in the Upper Volta. The concentrator had a diameter of 1.4 m and a focal length of 650 mm. The temperature reached 180°C at the focal point, the effective cooking power being 340 watts. The cooker had the capacity to hold 3 litres of foodstuff[32]. Results from the field study[33] showed that none of the cookers were used for cooking all the daily dishes because the cooking was slow. Although 40% of the women were reported to use the cooker regularly, some of the reasons given for not using the cookers included 'lack of sun', 'the pots were not suitable' and 'the quantity of food was too large for the solar unit'. A small intensive solar cooker survey in Sudan found that solar cookers were very useful for cooking small quantities of food or light foods[24]. Unlike the Upper Volta survey, in Sudan the traditional aluminium pots were very suitable for a solar cooker.

The technology for improving the fuel use of traditional woodfuel stoves is already available. There are also a variety of solar cooker designs. The main problems seem to be in the adoption of such units. Solar cookers still need to be cheaper. The type of cooker required much depends on local cooking habits and other sociological factors. However, it is unlikely that they can provide an adequate substitute for woodfuel, although they could certainly provide a complementary source of energy. With woodfuel stoves a potential 30% fuel saving could dramatically reduce demand. However, very often adoption depends on such factors as aesthetic appeal and even attributes of social status. Each community is unique with regard to cooking modes and each will pertain to different values with consequent perceptions of suitability. It is therefore difficult to generalize from one community to another as to the most appropriate and acceptable fuelwood stove or solar cooker.

Biogas

Biogas is a flammable gas produced by microbes when organic materials, such as aquatic weeds or the organic wastes in sewage systems, are fermented anaerobically within a particular range of temperatures, moisture contents and acidities. It can be used as a high quality fuel for cooking, lighting and power production. In chemical composition it is a mixture of some 60-70% methane with carbon dioxide and traces of other gases. The People's Republic of China has been recognised as the world's major user of biogas systems for many years and biogas production has become a comprehensive controlled method of waste disposal, supplying fertilizer and improving rural health in addition to providing a renewable energy source. By 1983 there were between 5 and 6 million operating biogas units and about sixty special biogas institutes in the provinces[34]. Their sole function is to carry out research and development work on biogas units and also to develop a trained group of technical teachers who would go out into the countryside to instruct others in the construction techniques, operation and maintenance.

In the province of Zhejiang, situated south of Shanghai, there are an estimated 370,000 units, of which 738 are located in an agricultural and fishing commune in the eastern outskirts of Hangzhou. The commune has over 4000 families and most of the workers are engaged in mixed farming, with vegetable production the main activity and a small number of sheep, cattle and pigs. The comparatively small number of biogas units in the commune is due to the limited volume of wastes and restrictions on the land available for housing. The basic family biogas unit is shown in Figure 6.6.

It is formed from a horizontal concrete cylinder, buried about one metre underground. Square-sectioned vertical entrance and exit chambers have tight fitting concrete lids. The gas is generated in the upper section of the cylinder and the delivery pipe to the family kitchen branches to a large vertical water manometer, mounted on the kitchen wall, so that a careful check can be kept on the gas pressure, normally about 250 mm of water above atmospheric pressure. The output during the summer months is approximately 6 to 7 m^3 of gas daily. Most families in the commune manage to keep their systems producing gas for at least six months each year, while the best units have given as much as ten months operation in a good year. Both human and animal wastes are used as raw material for the units, as well as various types of vegetable waste matter. Basic loading and clearing the processed waste for use as fertiliser takes between one and two hours per week in the summer months and slightly longer in the spring and autumn, as more care has to be taken with the quality of the wastes in colder conditions[34].

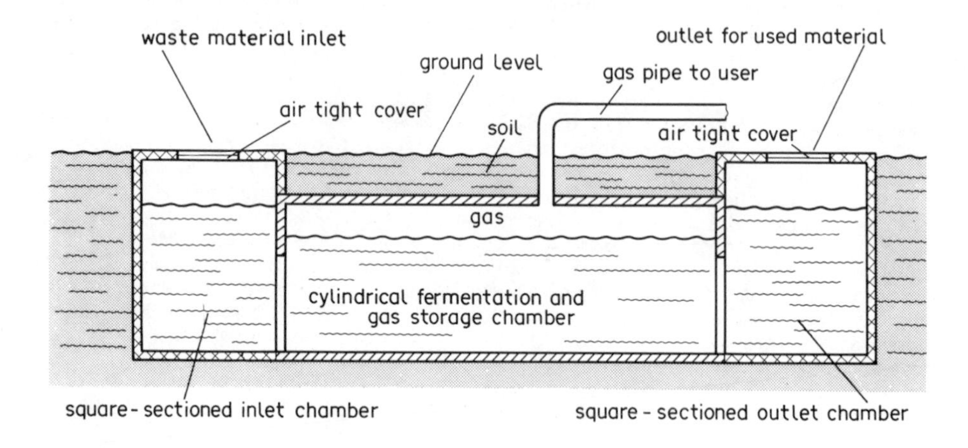

Figure 6.6

India has also had considerable experience with the development of biogas systems and some countries are now basing their designs on the established Chinese and Indian systems. For example over 1000 gobar (cow-dung) plants have been built in Nepal, biogas has been used to replace diesel fuel in Botswana and plans to develop some 300,000 biogas systems in rural Thailand have been studied[35]. Among the industrial countries Romania, with extensive pig farms, is using biogas in the experimental bus, shown in Figure 6.7.

Figure 6.7

In the UK, the temperate climate is less encouraging for biogas, but careful overall system design can overcome this problem. A prototype unit for a dairy herd of 320 cows was completed in Kent in 1979. Electrical power was generated from a Ford diesel generator modified for gas combustion with spark ignition, with a continuous maximum power calculated to be about 25kW[36]. A number of smaller units were also operating in other parts of the country[37].

A Fuel Alcohol Plant

An important factor in considering energy crop conversion is the energy needed for harvesting and for fertilisers to increase the crop yields. Net energy analysis is used to assess the energy cost/benefit ratio of any proposed fuel conversion process. The energy inputs and outputs of the system can be measured and the net energy ratio (NER) can be defined as the ratio of the energy outputs to the energy inputs. Any application of this concept requires a careful definition of the system boundaries. The NER concept has been used in the world's first cassava (mandioca) fuel alcohol commercial plant in Brazil[38]. The system boundaries and energy flows are shown in Figure. 6.8. The system consists of the cassava plantation, the fuel alcohol distillery and the forest from which the fuelwood is obtained to provide process steam for the distillery. Energy optimization of cassava distilleries could lead to the development of varieties of cassava with larger stalk-to-root ratios, so that the cassava stalks could replace the fuelwood requirement.

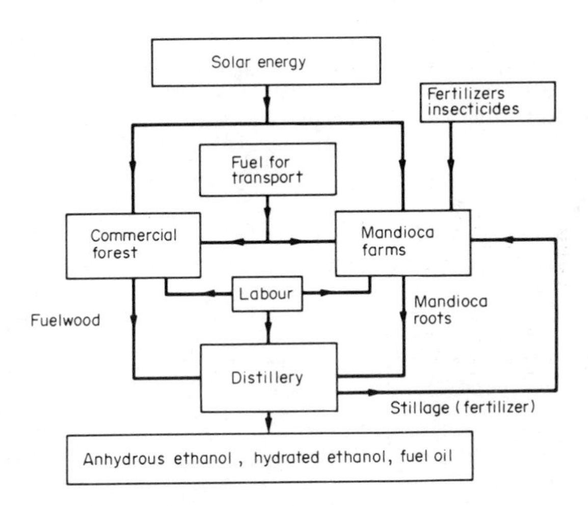

Figure 6.8

Another self-sufficient process is the sugar cane fuel alcohol system. The bagasse or by-product can generate all the necessary process steam. The NER of both systems is shown in Table 6.3. and based on 1 m^3 of anhydrous ethanol and total on-site generation of electric power.

Table 6.3

Raw Material	Output	Energy (10^6 kcal)				NER
		Input				
		Agriculture	Distillery	Transport	Total	
Sugar cane	5.59	0.42	0.017	0.26	0.70	8.0
Cassava	5.59	0.30	0.045	0.27	0.62	9.0

Data from Rio de Janeiro[38]

Short Rotation Forestry

The use of trees as energy crops has been proposed in several countries since the early 1970s. Detailed feasibility studies in the USA have shown that biofuels can be produced at competitive costs, by choosing the appropriate plant species, planting density and harvest schedule for each plantation site, thus minimizing the overall cost of the plant material[39]. In Ireland about 6% of the land area consists of bogland and less than a fifth of this area is being harvested for peat, which is either used directly as fuel in the home or for generating electricity, as discussed in Chapter 4. Until recently it had been thought that bogland was unproductive, but grass, shrubs and trees have all been successfully grown. Even with a conversion efficiency of 0.5% for Sikta spruce, the same bogland area at present used for turf could produce exactly half the quantity of electricity through the combustion of the trees. The Irish government has demonstrated that woodchips obtained from short rotation forestry can provide an economic alternative to oil[40]. Briefly, the short rotation forestry concept follows the sequence of selecting, planting, harvesting and utilizing as fuel the woodchips obtained from coppicing hardwood trees. The chips will be left to dry naturally in the fields, then collected, transported and burnt directly in specially modified power stations. Alternatively they can be bagged and sold directly for burning in central heating plants or for gasification. Harvesting would occur every three or four years. The trees are expected to regrow up to eight times from the existing root structure before replanting is necessary. A major advantage of the system is that the fuel can be stored indefinitely[40].

Greenhouse Applications

Full use is made of the available solar energy in a greenhouse design specially developed for colder regions at the Brace Research Institute[41]. It is oriented on an east-west axis with a large transparent south-facing roof. The rear, inclined north-facing wall is insulated with a reflective cover on the inner face. Heating requirements were reduced by up to 40% compared with a standard, double layered plastic covered greenhouse and increased yields of tomato and lettuce crops were reported. An ingenious method of passive energy collection and storage for greenhouses using the thermosyphon principle has been investigated in Australia[42]. During the day energy can be extracted from the air in a greenhouse atmosphere and transferred to the shallow soil by passing the air through a series of vertical pipes. The lower soil temperature will produce sensible heat transfer and some moisture will condense and pass into the soil. The passive system consists of a series of vertical holes in the ground, each with a concentric, smaller diameter pipe rising to a height of about 1.5 m above ground level. In daytime, this pipe is heated by solar radiation and produces an updraft, which in turn draws the moist air into the hole. The method was considered to be particularly worthwhile in dry areas with a wide range of ambient temperature from day to night and a shortage of water.

Both active and passive systems for heating the soil in greenhouses have been studied for many years in the Soviet Union and this work has been extensively reported in their literature. The following three examples represent a small proportion of their current programmes. Yakubov *et al*[43] have suggested that at a latitude of 40°, between October and March the maximum radiation occurs when the collector is orientated to the south and the angle to the horizon is 52°. Their shed-type greenhouse has a South glazed wall inclined at 52° and a long semitransparent North wall at 27° to the horizontal. This reduces heat losses and improves shading for the sowing area of the greenhouse. Solar energy is stored in the soil by a forced circulation system which sends the hot air through channels some 300 mm deep. At night the stored heat is gently released. The microclimate of a solar greenhouse with subsoil heating has been described[44] for a greenhouse with an enclosure coefficient of 1.53, defined as ratio of enclosure surface to floor area. The soil heating pipes were 150 mm in diameter and laid at a depth of 500 mm, 800-1000 mm apart. Studies were also carried out in a 50-30° greenhouse using forced electrically heated air circulation through the channels at night and on cloudy days. In this greenhouse on a typical January day at latitude 38° 50', an overall collection efficiency of 45% was recorded, based on the soil storage temperature rise[45].

Marine and Aqueous Applications

In oceans the production of organic matter by photosynthesis is generally limited by the availability of nutrients and they have been compared to deserts because of their low productivity. However, there are a few areas where natural flows bring the nutrients from the bottom of the ocean to the surface so that photosynthesis can take place. Particular interest has been shown in the cultivation of giant kelp *(Macrocystis pyrifera)*, a large brown seaweed found off the west coast of the USA. An early estimate examined the yield from an area of 600 000 km^2 and concluded that the equivalent of some 2% of the US energy supply could be provided[46]. One of the disadvantages of harvesting natural kelp beds would be the relatively low output caused by the lack of nutrients. Artificial kelp 'farms' have been suggested[47] and a 1000 m^2 system, with nutrient-rich deep water being pumped from deep water to the surface kelp, has been developed, as the first stage of a project which could lead to a 40 000 ha system.

Aquatic weeds can easily be converted to biogas. The water hyacinth *(Eichhornia crassipes)* has been extensively studied as a biogas source, particularly at the United States National Aeronautics and Space Administration (NASA)[48]. On a dry weight basis, one kilogramme of water hyacinth can produce 0.4 m^3 of biogas with a calorific value of 22 MJ m^{-3}. Aquatic weeds grow prolifically in many tropical regions and are costly to harvest. However, as weed clearance is essential to keep waterways clear, biogas production could be regarded as a valuable by-product in these applications.

Choice of System

The factors affecting the choice of a particular biological conversion system identified by Hall and Coombs[49] include considerations of agricultural capacity, environmental factors, population density, labour intensity in the agricultural sector and the energy demand per capita. Four regimes were distinguished in their simple classification:

(a) Temperate industrial areas such as North America, Western Europe and Japan where biomass will only produce a small fraction of the energy demand. Emphasis will be placed on the production of scarce chemicals from biomass. The by-products from certain industries may be used to provide heat and power, while the use of wood as a direct fuel source is also possible.

(b) Tropical and sub-tropical regions with good soil and high rainfall such as parts of India and Africa, Brazil, Indo-China and north Australia. Energy from biomass has the greatest potential in these regions, with many examples already competing economically, e.g. biogas and the more efficient use of fast growing wood species.

(c) Northern polar and arid regions where biological systems are only possible in an artificial environment, e.g. use of the nutrient film technique.

(d) Marine and aqueous regions through the use of fast growing water weeds, seaweeds or micro-algae.

The development of photobiological energy conversion systems can take place more readily in the temperate western countries with their high technological background. However, these systems can function more effectively in the developing tropical and sub-tropical countries and could make a very significant contribution towards reducing their dependence on increasingly scarce and expensive oil.

Summary

Hydropower is the only renewable energy resource with a fully developed technological base and a very predictable growth rate over the next few decades. Its industrial infrastructure is well established in many countries and it provides very substantial proportions of the electricity demand in a number of countries. Although it accounted for only 6.5% of the world's primary energy consumption in 1982, this figure could easily rise to over 10% in the early part of the next century. It is particularly suitable for the needs of remote communities in the developing countries.

Biomass is already the most important energy resource for between a quarter and half of the world's population. Its use on a commercial scale is increasing through technical development. Further increases in the costs of competitive energy will bring many more applications into widespread use. The resource base is very large and capable of at least a fourfold expansion from today's level which is estimated to lie between 6 and 13% of total world energy use. There would appear to be no major technical obstacle to a fourfold expansion.

References

1. Foley, Gerald, The Energy Question, Second Edition, Penguin Books Ltd., 1981.

2. Electricity Supply in Great Britain, The Electricity Council, London, 1973.

3. A History of Technology, Oxford University Press, V, 528-533, 1962.

4. Deudney, Daniel, Rivers of Energy: The Hydropower Potential, Worldwatch Paper 44, Worldwatch Institute, Washington, 1981.

5. Power Resources of the World, 1929.

6. Survey of Energy Resources, World Energy Conference, London, 1980.

7. BP Statistical Review of World Energy 1982, The British Petroleum Company plc, London, 1983.

8. Report of the Technical Panel on Hydropower, United Nations Conference on New and Renewable Sources of Energy, Nairobi, 1981.

9. BP Statistical Review of the World Oil Industry 1975, The British Petroleum Company, London, 1976.

10. Annual Report and Accounts 1982-83, Central Electricity Generating Board, London, 1983.

11. Knight, P. Report from Brazil, Electrical Review International 1, 1, 1984.

12. Acker, F., Waiting for hydro to plug power shortfall, Ibid.

13. Hall, D.O., Solar energy use through biology - past, present and future, Solar Energy, 22, 307-329, 1979.

14. Energy in the Developing Countries, World Bank, Washington, D.C., 1980.

15. Brandt, Willy (Chairman), North-South: A programme for survival, The Report of the Independent Commission on International Development Issues, Pan Books, London and Sydney, 1980.

16. Makhijani, A. and Poole, A., 'Energy and Agriculture in the Third World', Ballinger Publishing Co., Cambridge, Massachusetts, 1975.

17. Jesch, L.F., Solar Energy Today, UK Section, International Solar Energy Society, London, 1981.

18. Ward, R.F., Alcohols as fuels - the global picture, Solar Energy, 26, 169-173, 1981.

19. Hancock, G., Premiers to Discuss Use of Renewable Energy, The Guardian, 10th August, 1981.

20. Harrison, P., Inside the Third World - The Anatomy of Poverty, Second Edition, Penguin Books Ltd., 1981.

21. 'Domestic Energy in Sub-Saharan Africa: The Impending Crisis: Its Measurement and the Framework for Practical Solutions', F.R.I.D.A, December, 1980.

22. Singer, H., 'Report to the Government of Indonesia on Improvement of Fuelwood Cooking Stoves and Economy in Fuelwood', Food and Agricultural Organization, Rome, 1961.

23. 'Wood Conserving Cookstoves - A Design Guide', V.I.T.A./I.T.D.G., 1980.

24. Brattle, L., 'Novel and Improved Cookstove Technology For Use in The Sudan: The Application of Home Economics to the Question of Appropriate Technologies', Ph.D. Thesis, University of Surrey, July 1983.

25. Duffie, J.A., Lof, G.O.G. and Beck, B., 'Laboratory and Field Studies of Plastic Reflector Solar Cookers', New Sources of Energy - Proceedings of the U.N. Conference, 5, Paper S/87, 1961.

26. Von Oppen, M., 'The Sun Basket', Appropriate Technology, 4, 3, 7-10, 1977.

27. Tabor, H., 'A Solar Cooker for Developing Countries', Solar Energy, 10, 4, 153-157, 1966.

28. Hoda, M.M., 'Solar Cookers', International Conference and Exhibition of Solar Building Technology, Vol II, Paper 8/10, July 1977.

29. Telkes, M. and Andrassy, S., 'Practical Solar Cooking Ovens', New Sources of Energy - Proceedings of the U.N. Conference, 5. Paper S/101, Rome, 1961.

30. Bernard, R., 'Easy to Build Solar Cookers', International Conference and Exhibition on Solar Building Technology, Vol II, Paper 8/11, July 1977.

31. Alward, R., 'Solar Steam Cooker', Brace Research Institute, (Do it yourself leaflet L-2) October 1972.

32. 'Solar Cookers - Comparison Between Different Systems of Solar Cookers Considering Both Technical and Economic Aspects', G.A.T.E., 1978.

33. Oedrago, O. Dieudonne, 'Enquete D'Evaluation des Cuisinieres Solaires', Centre National de la Recherche Scientifique et Technologique, France.

34. McVeigh, J.C., Personal discussions in Hangzhou, People's Republic of China, March 1983.

35. RERIC News, Renewable Energy Resources Information Centre, P.O. Box 2754, Bangkok, Thailand, 4, 3, December 1981.

36. Keable, J. and Dodson, C.A., Modular System for Biogas Production Using Farm Waste, Sun II, 1, 83-87, Pergamon Press, Oxford, 1979.

37. Chesshire, Michael, Anaerobic Digestion of Farm Wastes, in Solar Energy in Agriculture, Conf. C.33, UK-ISES, London, 1983.

38. Centro de Tecnologia Promon Newsletter, 3, 1, Rio de Janeiro, February 1978.

39. Szego, G.C. and Kemp, G.C., Energy Forests and Fuel Plantations, Chemtech, 275-284, May 1973.

40. Neenon, M., Lyons, G. and O'Brien, T.C., Short Rotation Forestry as a Source of Energy, Solar World Forum, Vol. 2., 1258-1262, Pergamon Press, Oxford, 1982.

41. Lawand, T.A., Alward, R., Saulnier, B. and Brunet, E., The development and testing of an environmentally designed greenhouse for colder regions, Solar Energy, 17, 307-312, 1975.

42. Morrison, G.L., Passive energy storage in greenhouse, Solar Energy, 25, 365-372, 1980.

43. Yakubov, Yu. N., Shodiev, O. Kh. and Imomkulov, A., A shed-type hothouse with solar energy accumulated by soil, Geliotekhnika, 15, 50-53, 1979.

44. Vardiyashvili, A.B. and Khairitdinov, B., Formation of microclimate in a solar greenhouse, Geliotekhnika, 15, 87-91, 1979.

45. Vardiyashvili, A.B. and Khairitdinov, B., Heat exchange in a solar greenhouse with a soil heating system, Geliotekhnika, 15, 3, 68-72, 1979.

46. Wolf, M., Utilization of solar energy by bioconversion - an overview, Testimony before the US House of Representatives Science and Astronautics Committee, 13 June 1974.

47. White, L.P. and Plaskett, L.G., Biomass as Fuel, Academic Press, London, 1981.

48. Making Aquatic Weeds Useful: Some perspectives for developing countries. National Academy of Sciences, Washington,1976.

49. Hall, D.O. and Coombs, J., The prospect of a biological-photochemical approach for the utilization of solar energy, pp. 2-14 in Energy from the biomass, The Watt Committee on Energy, Report No. 5, London, 1979.

Further Reading

The literature on hydropower is so vast that the United Nations Technical Panel[8] could not find it appropriate to present a list that could do justice to the technology. The basic theory is well covered in many standard undergraduate texts. A good starting point for detailed technical study is Hydroelectricity Prospects in the New Energy Situation, proceedings of a symposium of the United Nations Economic Commission for Europe held in 1979 and published by Pergamon Press in 1981. A non-technical survey of hydropower can be found in David Deudney's 'Rivers of Energy : The Hydropower Potential, Worldwatch Paper 44, The Worldwatch Institute, Washington, 1981.

Biomass has a relatively smaller, but expanding literature. The key reference is Professor David Hall's review paper Solar Energy Use Through Biology - Past, Present and Future, published in Solar Energy, 22, 307-328, 1979. This contains nearly 150 other reference sources. Hall is also joint author with G.W. Barnard and P.A. Moss of the Pergamon Press book Biomass for Energy in the Developing Countries, published in 1982, which contains good data from national surveys and lists of research groups and further reference sources. Two useful introductory books are Biological Energy Resources, by Malcolm Slesser and Chris Lewis, published by E. and F.N. Spon Ltd., London in 1979 and Biomass as Fuel by L.P. White and L.G. Plaskett, published by Academic Press in 1981.

Exercises

1. In the Dinorwig pumped storage hydroelectric power station the mean height of the Marchlyn reservoir is 568m above the power station. If the overall conversion efficiency is 86% and the plant provides 1620 MW for four hours, how much water has flowed to the lower Peris reservoir in this time?

2. Examine developments in the Mahaweli hydropower project in Sri Lanka. How could the country's economy be altered when the project is completed? What other indigenous sources of energy could be harnessed in Sri Lanka?

3. The Cahora Bassa hydropower scheme on the Zambesi River in Mozambique was unable to supply its only customer, South Africa, during the last few months of 1983 because the transmission lines had been destroyed by guerilla action. Discuss the implications of this type of disruption on international cooperation in the development of world hydropower resources.

4. Why can the annual consumption of hydroelectricity be plotted directly on a logistic curve? Derive a logistic equation for hydropower using the data given in Table 6.2. and assuming the economically available potential is 50% greater than the values shown in Table 6.1.

5. One of the problems associated with hydropower management is the silting of the reservoir. What measures can be taken to reduce or eliminate this problem? The Aswan High Dam has greatly reduced the supply of silt to the Nile Delta. What effect has this had on the agriculture?

6. Examine the energy needs of a rural community of about 2000 people in a temperate climate such as North West Europe. What area of forest would be needed under short rotation forestry to satisfy the energy demands of this community? How much employment would it create?

7. Like many other countries which had major forest resources some 30 years ago the People's Republic of China experienced a period of rapid increase in wood consumption without an associated programme of replacement. What steps has the Chinese government taken to reverse the deforestation?

8. Examine the possibility of making a 500 hectare mixed farm in Northern Europe completely self-sufficient in energy.

9. The conversion of biomass to methanol is already becoming competitive with other conversion routes in some countries. What is the present state-of-the-art in methanol production from biomass?

10. To what extent could the developments in greenhouse heating reported from the USSR be adopted in other countries? What advances in greenhouse heating using renewable energy resources have occured recently?

Answer

1. $4.868 \times 10^6 m^3$

7. NUCLEAR POWER

Introduction

Radioactivity was discovered by Becquerel in Paris in 1896 while examining the fluorescent properties of uranium[1]. In the following year J.J. Thomson proved the existence of the electron at the Cavendish Laboratory, Cambridge. Madame Marie Curie showed that thorium possessed similar properties and found a new radioactive element, radium, in pitchblende, one of the natural ores of uranium. By 1902 Rutherford and Soddy[1] had decided that

> ...the atom of the chemist, although still the ultimate limit of subdivision of matter in every artificial engendered process, is not the natural limit.

Rutherford went on to announce the transmutation of nitrogen in 1919. Two years later Soddy[2] was able to point out that studies of radioactivity and of the internal structure of the atom had proved that there were in ordinary materials

> ...amounts of energy of the order of a million times that which could be obtained from fuel during combustion.

In 1932 Cockcroft and Walton became the first to split the atom, confirming the early predictions of Rutherford and Soddy, and Chadwick identified the neutron, the nuclear particle with no charge which was to play a vital role later in chain reactions[3]. Hahn and Strassmann discovered fission in 1938 and the first steps to obtain the controlled release of the energy in the nucleus of heavy atoms had been taken.

By the spring of 1941 a group of British nuclear physicists reported on the feasibility of using nuclear energy as a source of power. This information was shared with the United States where, in November 1942, the first nuclear reactor was constructed under the direction of Enrico Fermi in a disused squash court at the University of Chicago. It was started on the 2nd December 1942, with the rate of heat generated being limited to about 0.5 watts[4]. From that point priority was given to the 'Manhattan Project', the development of the atomic bomb, and a substantial number of British scientists worked with the Americans on this project[3]. Others went to Canada to work with a heavy water moderated reactor. At the end of the war in 1945 the Americans took steps to restrict access to information about nuclear energy. The McMahon Act, passed in 1946, established the United States Atomic Energy

Commission (AEC) and completely excluded the British and Canadian scientists and engineers from the nuclear information for which they had been jointly responsible with the Americans.

The next few years saw increasing efforts by the United States, the USSR and the UK to develop their nuclear weapons programmes. At the same time some studies indicated that the generation of electricity from nuclear reactors could become less expensive than electricity generated from coal[4]. In 1954 the Chairman of the United States Atomic Energy Commission, Lewis L. Strauss, predicted that nuclear electricity would be produced so cheaply and in such quantities that it would not have to be metered[5]. People would be able to use as much as they wanted for a low monthly charge. Another major issue was safety. The safety of nuclear power reactors was discussed at the first Geneva Conference on the Peaceful Uses of Atomic Energy held in 1955. This was followed by several theoretical studies estimating the likelihood of a major disaster. The most pessimistic prediction from the United States[6] suggested that this was no worse than 1 in 100,000 per reactor year of operation. Although this theoretical major disaster was clearly very unlikely, there was widespread concern at the projected cost of possible insurance claims. The resources of the large private insurance groups and the electricity utilities would be unable to cover such an accident. To overcome this problem, the United States Houses of Congress passed the Price-Anderson Act in 1957. This act instructed the utilities to obtain as much private insurance cover as possible, with the government providing a further 500 million dollars. At that time the private cover came to 60 million dollars. The act limited the total financial liability to 560 million dollars, a sum which was less than a tenth of the maximum 'worst case' claim[6].

In the United Kingdom the world's first 'commercial' nuclear power station, Calder Hall, was opened on the 17th October 1956 at Windscale. However, it is not widely appreciated that the prime purpose of the Calder Hall reactors and four other small reactors built at Chapelcross in Scotland was to produce plutonium for weapons. The first group of large scale civil British reactors developed from the Calder Hall design. Known as Magnox reactors, their name was taken from the magnesium alloy used as the fuel cladding material. For a number of years the UK were world leaders in civil nuclear power, generating more than half the total world output annually from 1956 to 1968. But from this point her relative position declined. Over the next ten years the world nuclear electricity programme expanded very rapidly with a compound annual growth rate of about 25%. This is shown in Figure 7.1 which gives the primary energy equivalent of nuclear electricity generated in the world and the UK since 1968[7], together with predicated growth patterns.

By 1974 the worldwide installed capacity, including smaller experimental reactors, was over 70,000 MWe. There were about 150 reactors in operation with a further 250 under construction, ordered or planned. The International Atomic Energy Agency predicted that by 1980 the worldwide capacity would be over 25,0000 MWe - a straightforward continuation of the 25% per annum growth rate seen in the previous six years. Only half this figure was achieved. The major official nuclear bodies in the United States, the UK, France, and organisations such as the Organisation for Economic Cooperation and Development and the European Economic Community predicted that 50% or more of the world's electricity demand would be met by nuclear energy by the year 2000[4].

One prediction in 1972 suggested that the United States alone would have installed 450,000 MWe of nuclear power by 1985[8].

Recalling the period up to the early 1970's, David Freeman[9], Managing Director of the Tennessee Valley Authority in the United States, commented in 1983

> This was the new technology. Everyone believed in it. It had an aura of scientific credibility and the support of Congress...

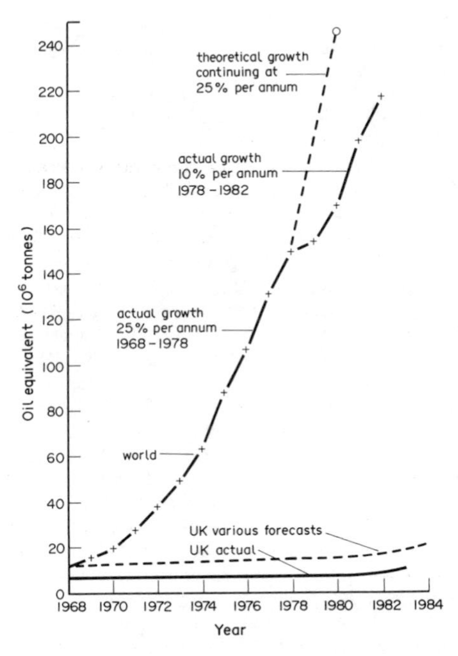

Figure 7.1 World and UK nuclear energy consumption

But already major technical problems were apparent. Both in the United States and in the UK the move to larger, relatively untried reactor systems resulted in costly delays in construction and breakdowns in operation. In 1973 the Wall Street Journal noted that one of the most dependable features of a nuclear plant was its unreliability, with breakdowns blamed on faulty engineering, defective equipment and operating errors[10]. The price rises in oil during 1973 and 1974 which should have been advantageous for nuclear power caused the world economic recession and, together with conservation measures, slowed the increase in the demand for electricity. In the four year period from 1975 to 1978 there were only 13 new reactor orders and some earlier orders were cancelled or postponed. A number of independent consultants, such as Lovins[11] and Foley[12], drew attention to the way in which predictions of installed nuclear capacity, based on the 25% per annum growth rates of the late 1960s and early 1970s, had been falling every year since 1970. For example projections of 1985 Nuclear Generating Capacity for the O.E.C.D. countries had fallen from 563,000 MWe in 1970 to 214,000 MWe by 1977.

However, the research scientists and engineers within the industry were well aware that some of the technical problems had not been fully solved. For example, they knew during the 1970s that vibrations within the reactor cooling system were caused by high velocity coolant flows. The sequence of events in a typical problem situation have been clearly described by Blevins[13]. If a reactor operator observes something unusual such as excessive temperature levels or vibrations, he reduces the power output. The reactor owners, constructors and governing safety board are informed. The analyst is then asked to diagnose and correct the problem under severe safety and economic restraints. Every hour that a reactor is not generating electricity is a net loss to the operator. Blevin's comment is perceptive

> This is certainly not the atmosphere to encourage leisurely scientific work covering a spectrum of test conditions. Moreover, once a problem is resolved, it still may not be known in quantitative detail just how the solution worked, and the analyst may not be permitted to publish his experiences.

Blevins submitted his paper for publication in February 1979. Two weeks later the International Herald Tribune[14] reported that many Americans had either lost or were beginning to lose confidence in assurances that nuclear energy was safe. Both government and the nuclear industry were considered to have been too careless in their safety studies, too committed as advocates of nuclear energy, occasionally deceptive or misleading in that advocacy and consistently over-optimistic both in their safety estimates and their judgement of the extent to which the public would accept questionable assurances. Within four weeks the most serious and most expensive accident in the history of the nuclear industry occurred. This was the failure of the Three Mile Island Unit 2 Pressurised Water Reactor in the United States on the 28th March 1979 (described in more detail on page 152). The effect of this accident is clearly seen in Figure 7.1. Many countries closed down their PWRs for modifications and the fitting of additional safety control equipment. The percentage increase in nuclear generated electricity over the next year, some 4%, was the lowest recorded over the 15 year period from 1968 to 1983. In the UK the actual nuclear output has lagged behind the official forecasts for broadly similar reasons to those in other countries - delays in construction and breakdowns in operation. Although by 1983 the average annual growth rate in world nuclear generated electricity had fallen to 10% since 1978, this was nevertheless a very high figure, representing a doubling time of just over 7 years. Reactor performance is assessed by the load factor. This compares the electricity supplied during a period of time with the theoretical amount which could have been supplied if the reactor were operating continuously at its maximum rated output. Of the 223 reactors in the Western World, 47 had achieved load factors between 80% and 97% in the year up to 31st March 1983[15]. Over half were performing with load factors greater than 60%. At the bottom of the table nine reactors, including both Units at Three Mile Island, had not been operating during the year.

Some Basic Concepts

Nuclear energy is the energy released by a nuclear reaction or by radioactive decay. This energy is released during the regrouping of the particles which make up the nucleus of an atom. Nuclear power is obtained when this released energy is converted first into heat and then into electricity.

An atom is the smallest particle of any element which has the chemical properties of that element. The nucleus of an atom is not solid but consists of two different kinds of particles, the positively charged protons and the neutral particles called neutrons. The comparatively massive nucleus is always positively charged and is surrounded by negatively charged electrons which move in orbits around it. The net electrical charge of the atom is zero, so that for any particular atom the number of orbiting electrons is exactly equal to the number of protons. All the atoms of any particular chemical element contain the same number of protons. The number of protons is also the atomic number of the element. For example, hydrogen has only one proton in its nucleus and a single electron and its atomic number is 1. Uranium has 92 protons and 92 electrons and its atomic number is 92. The sum of the numbers of neutrons and protons gives the mass number and determines the mass of the nucleus.

The atomic number determines the chemical properties of an element. Because this depends only on the number of electrons, or protons, in the atom, some atoms which are chemically the same can have different masses, depending on the number of neutrons in the nucleus. Different types of chemically similar atoms with the same number of protons but with different numbers of neutrons are called isotopes. Hydrogen has two naturally occuring isotopes; normal hydrogen with one proton, mass number 1, and deuterium with one proton and one neutron, mass number 2. Uranium has three naturally occuring isotopes. Each has 92 protons, but with 142, 143 and 146 neutrons in their nuclei to give mass numbers of 234, 235 and 238 respectively. Elements are sometimes identified by their mass numbers e.g. a carbon atom with 6 protons and 6 neutrons is called carbon-12.

Hydrogen and elements with small mass numbers are known as 'light' atoms, while uranium and elements with mass numbers of two hundred and above are known as 'heavy' atoms.

Some of the heavy complex isotopes, such as uranium and radium are unstable and the nucleii of their atoms tend to split or transform into more stable atoms. These substances are radioactive and the transformation process is known as radioactive decay. During this process three main types of nuclear radiation are emitted - alpha particles, beta particles and gamma rays. Alpha particles are identical to those of the nucleus of the helium atom, consisting of two protons and two neutrons. Beta particles are negatively charged electrons. Gamma rays are electromagnetic radiation. Other types of this class of radiation are radio waves and visible light. The rate at which the different types of radioactive nucleii naturally decay, or change into a different element, varies enormously. A brief discussion of radioactive decay and the half-life was given in Chapter 1.

Nuclear Fission and Breeding

In 1938 two Germans, Otto Hahn and Fritz Strassmann, discovered that uranium-235 could interact with an additional neutron, absorbing it and dividing into two fragments of approximately equal mass, releasing a large amount of energy and emitting several neutrons. The mass of the two fragments is always less than the original mass. These missing or 'lost' fragments have been largely transformed into kinetic energy, which, in turn, is converted into heat as the fission products collide with the surrounding atoms and are brought to rest. A neutron emitted during the fission process can strike the nucleus of another uranium-235 atom, releasing more energy and emitting more neutrons. This continuous splitting process is called a chain reaction. It can be controlled in a nuclear reactor to produce a steady heat output. This fission process can only take place if the neutrons, which are emitted with high velocities, are slowed down within the reactor.

The energy produced through the complete fission of one kilogram of uranium-235 is theoretically equivalent to that released during the combustion of about 2000 tonnes of oil. Each fission releases only 3.2×10^{-11}J, but a kilogram of uranium-235 contains 2.5×10^{24} atoms, giving an energy source equivalent to 8×10^{13}J. Natural uranium contains 0.7% of uranium-235. The energy equivalent of one tonne of uranium burnt in a thermal power station depends on the type of reactor. In the UK, British Nuclear Fuels suggest a figure of 7,200 tonnes of oil for the Magnox reactors and 8,400 tonnes for the Advanced Gas Cooled Reactors.

The remaining 99.3% of natural uranium is uranium-238. This is not suitable for producing energy because it absorbs neutrons without undergoing fission. After absorbing a neutron, a process also known as neutron capture, uranium-238 becomes uranium-239, which decays very rapidly into plutonium-239, an element which does not occur naturally. Uranium-238 is known as a 'fertile' element, as the new element, plutonium-239 can be obtained from it. This is known as breeding. Plutonium-239 can also undergo fission, but only if it is bombarded by high velocity or fast neutrons.

Commercial Nuclear Reactors

Nuclear reactors are designed to produce heat that can be used to generate electricity, normally in a conventional thermal steam cycle. The primary function of the nuclear reactor is to achieve, sustain and control a chain reaction by the fission process. Because natural uranium contains only 0.7% of the fissile uranium-235, the chance of at least one of the few neutrons from every fission causing another fission is not high. Some thermal reactors, such as the Advanced Gas Cooled Reactor and the Pressurized Water Reactor, use 'enriched uranium' in which the uranium-235 content has been increased up to 2 or 3%. Although the enriching process is highly complex and energy intensive it has certain economic advantages. The core size can be reduced and the maintenance of the chain reaction is less critically dependent on geometrical arrangement. There are two ways of achieving fission. The first is by slowing down the fast moving neutrons by means of a moderator, a special material in the reactor core. In the second a sustainable fission chain reaction is achieved without a moderator, but with very high enrichment in excess of 20%. This is the basis of fast reactor design. In addition to the moderator and the reactor fuel core, the other main factors of a nuclear reactor are the primary coolant loops and pumps and the control rods.

The fuel core contains the fuel elements in fuel rods. The chain reaction of the fission process takes place in the core. The primary coolant loops extract the heat generated by fission and transfer it to the steam plant. Neutrons slowed down in the moderator are called 'thermal' neutrons, and reactors with moderators are called thermal reactors. The moderator is also a further line of defence in the reactor safety system. If the reactor temperature were to rise rapidly through an unplanned acceleration of the chain reaction, the moderator would be capable of acting as a huge heat sink, absorbing the extra heat from the reactor. The control rods are made of highly neutron absorbent materials such as cadmium or boron. Moving the control rods in or out of the reactor core controls the number of neutrons available for fission and the amount of heat generated. The fuel elements in the core must be replaced with fresh elements at regular intervals. Most reactors need to be completely shut down for this operation which reduces their potential for achieving very high load factors. A massive steel and concrete structure known as the containment structure surrounds the reactor assembly of fuel, moderator, control rods and coolant to contain the internal high pressure systems. This also protects the surrounding environment from any release of radioactivity in the event of a serious failure in the primary coolant circuit.

The basic arrangements of the fuel elements, moderator, coolant and containment for the four main thermal reactor systems are shown in Figure 7.2. A summary of some operating design parameters is given in Table 7.1.

The Magnox reactor, and its successor in the UK the Advanced Gas Cooled Reactor, are the only types to use graphite as the moderator and carbon dioxide gas as the coolant. The carbon dioxide transfers its heat to water in a steam generator. This steam then drives a turbine coupled to an electric generator. The fuel for Magnox reactors is uranium metal clad in a magnesium alloy (Magnox) with low neutron absorption.

The Boiling Water Reactor (BWR), developed in the United States, uses ordinary water as the moderator in a direct system in which the water boils under pressure and also acts as the coolant. The steam from the boiling coolant drives a turbine coupled to an electric generator. The fuel is uranium dioxide inside zirconium tubes, known as Zircaloy cans.

The Pressurized Water Reactor (PWR) was also developed in the United States and uses ordinary water as the moderator. This water is at very high pressure and is pumped to a steam generator where its heat is transferred to a secondary steam circuit. This steam drives a turbine coupled to an electric generator.

The Pressurized Heavy Water Reactor (PHWR) or CANDU reactor was developed in Canada (CANDU is derived from Canada, Deuterium and Uranium) and uses heavy

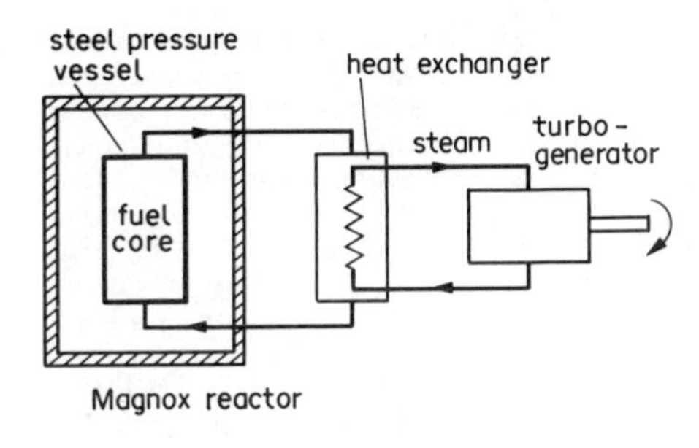

Magnox reactor

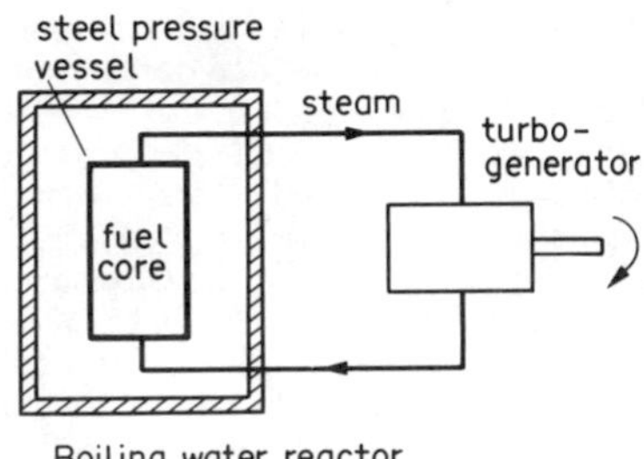

Boiling water reactor

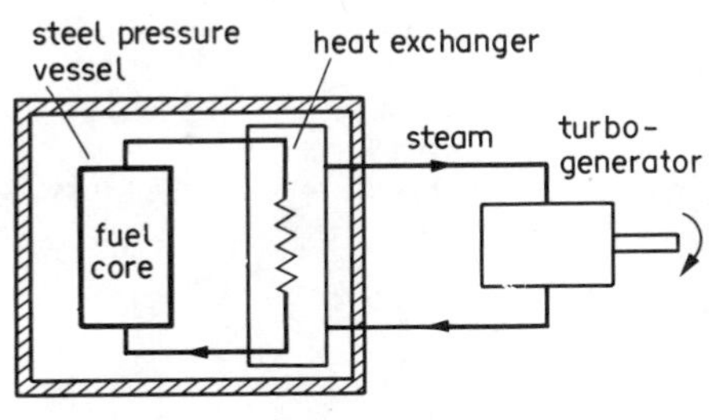

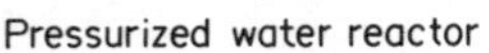
Pressurized water reactor

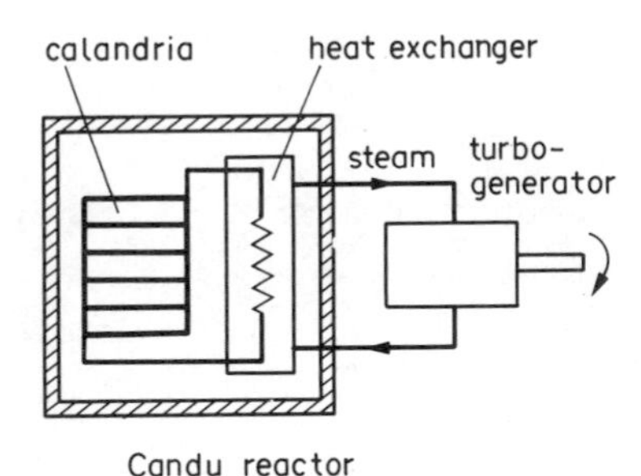

Candu reactor

Figure 7.2

water as the moderator and, in a separate pressurized circuit, as the coolant. Heavy water contains deuterium, whose nuclei consist of one proton and one neutron. The presence of the neutrons means that the heavy water can absorb no new neutrons but

Table 7.1.

Reactor	Magnox (UK) 600MWe	BWR (USA) 600MWe	PWR (USA) 700MWe	PHWR(CANDU) (CANADA) 600MWe
Uranium Enrichment % Uranium -235	0.7 (natural)	2.6	3.2	0.7 (natural)
Coolant outlet temperature (°C)	400	286	317	305
Coolant pressure (lbf/in^2 absolute)	300	1050	2235	1285
Steam cycle efficiency %	31	32	32	30
Core diameter (metres)	14	3.7	3	7.1
Core height (metres)	8	3.7	3.7	5.9

Indicative data from 'Nuclear Power Reactors', UKAEA and Nuclear Power Company Ltd., 1977.

slows down the fast neutrons from the uranium-235, acting as a very efficient moderator. Canada has large natural uranium resources and has not developed facilities for uranium enrichment. Natural uranium was selected as the fuel because heavy water was used as the moderator. The particularly interesting engineering feature of the CANDU reactors is that each cluster of the uranium dioxide fuel elements is placed in a separate pressure tube. The pressure tubes are in a large tank of heavy water, called the Calandria. The use of pressure tubes eliminates the need for a pressure vessel. If a failure occurs in an individual pressure tube, this is quickly detected and the tube is easily replaced. Fuelling can take place continuously on load. The attention paid by the Canadians to quality and integrity in their engineering applications has resulted in the life-time performance of the large CANDU reactors being easily the best in the world. This is shown in Table 7.2. which gives cummulative load factors of the six best large (greater than 500 MWe) reactors in the world up to 31st March 1983.

Table 7.2

Place	Reactor	Country	Type	Cumulative Load Factor %	MWe gross
1st	Bruce 3	Canada	CANDU	87.2	826
2nd	Bruce 4	Canada	CANDU	84.5	826
3rd	Pickering 2	Canada	CANDU	82.8	542
4th	Stade 1	FR Germany	PWR	82.3	662
5th	Pickering 4	Canada	CANDU	81.0	542
6th	Bruce 1	Canada	CANDU	80.8	826

Derived data from Nuclear Engineering International[15].

The average annual load factors for each of the four main systems are shown in Table 7.3. As the range of reactor output now extends from 150 MWe to 1300 MWe, the compilers of the Table 7.3 have suggested that a weighted average against output gives a better representative figure for the system. When the weighted figures are compared with the unweighted, the slightly lower figure for the PWR reflects the modifications to a number of reactors which were carried out during the year. The higher figures for the other three systems indicate a superior performance from the higher output stations.

Table 7.3

Annual Load Factors
Year ended 31st March 1983

System and number of reactors	Average %	Weighted Average %	Total reactor output, MWe
CANDU (13)	70.9	79.7	6664
PWR (104)	60.4	59.9	85909
BWR (56)	60.3	61.5	41559
Magnox (26)	57.0	57.8	8527

Data from Nuclear Engineering International[15].

Cumulative figures include the complete history of the nuclear power station's generating achievement and are given in Table 7.4. The cumulative figures for the Magnox stations are greater than the annual figure, as a number of Magnox reactors were shut down for repairs during the year.

Table 7.4

Cumulative Load Factors up to 31st March 1983

System and number of reactors	Average %	Weighted Average %	Total reactor output MWe
Magnox (26)	60.2	57.9	8527
CANDU (17)	59.7	63.8	9247
PWR (112)	56.4	54.7	93428
BWR (59)	55.2	54.5	44766

Data from Nuclear Engineering International[15].

The much higher annual figures for the other reactors also reflect the more recent starting-up periods for some of these systems, when full rated outputs take some years to achieve. Several CANDU systems had operated for less than a year by 31st March 1983.

National figures for the major Western countries are given in Table 7.5. These include the British Advanced Gas Cooled Reactors which are excluded from the main analysis because sufficient operating expensive is not yet available. The performance of the UK reactors should improve during 1983-4 as the Magnox stations return to full operation and the new Advanced Gas Cooled Reactors pass through their starting-up phase. Canada should also show an improved annual performance for the same reason, as a result of several CANDU reactors commissioned in 1982.

Table 7.5

National Performances up to 31st December 1982

Country	Average annual load factor %	Number and output of reactors MWe	Average cumulative load factor %	Number and output of reactors MWe
Canada	77.1	10 (5959.6)	55.2	13 (7864.6)
France	53.0	28 (22956)	52.9	30 (24826)
FR Germany	68.0	11 (10248)	60.1	11 (10248)
Japan	65.8	24 (16776)	57.9	25 (17432)
Sweden	68.3	9 (6787)	53.2	10 (7767)
UK	51.4	23 (8219.3)	52.5	23 (8219.3)
USA	55.3	73 (61073)	53.4	77 (65430)

Data from Nuclear Engineering International[15].

Some Economic, Social and Political Issues

It is very difficult to present a balanced objective view of all the issues relating to nuclear power. The well-known scientific method approach used by engineers and scientists for developing solutions to complex problems[16] can only be applied to some of the nuclear issues with caution. The first steps in scientific problem solving are to recognise the problem and to define it. This is followed by the gathering and compilation of preparatory material in the form of data, ideas, opinions and assumptions. With some issues relating to nuclear power the data is not available. It is not just unknown. It is unknowable with today's technology. Opinions and assumptions are scattered throughout the literature. It is often impossible to say definitively that any one view is correct. In the following sections only one issue, that of the cost of nuclear electricity, is explored in depth. In this case it is possible to pursue some of the facts behind reasons why the industry has been stating that nuclear electricity is cheap and critics state the opposite.

The Cost of Nuclear Electricity

Although this would appear to be an issue which can be examined on the basis of published estimates and actual peformance figures, this has not proved to be straight forward. The main difficulty is that a nuclear power station is relatively expensive to build. Should a station take longer to build than the estimated time, the capital costs rise. This happened with the Advanced Cooled Reactor programme in the UK. Based on figures provided by Professor Duncan Burn for the House of Commons Select Committee on Energy in 1980, Sweet[17] has calculated a full economic cost of six to seven times the original estimate after allowing for inflation.

By the end of 1983 Dungeness B was not fully commissioned over 17 years after construction commenced on site. However, the first unit at Torness, which was started in 1980, was reported to be on target for commissioning in 1987[18]. Once built it is claimed that its fuel and operating costs will be much lower than those of a corresponding coal-fired power station, which costs perhaps half as much to build[19]. How should a large capital cost be repaid over the lifetime of a power station? Even without the added complication of inflation the answer depends on a number of assumptions. For example the equation used to calculate the constant annual payment P necessary to repay a capital loan C in n years at a fixed annual interest rate r is

$$P = \frac{Cr\,(1 + r)^n}{(1 + r)^n - 1} \tag{7.1}$$

Capital costs of power stations are often expressed as a unit cost per kilowatt hour of design capacity.

Example 7.1. The capital cost of a nuclear power station is £1500 per K.W. (very close to estimates for the Sizewell PWR in the UK). Two different sets of operating conditions are to be examined: (a) The repayment period is 25 years with an annual interest rate of 5%. The station operates with a load factor of 75% and (b) the repayment period is 20 years with an annual interest rate of 10%. The station operates with a load factor of 55%. Compare the cost of repaying the capital loan on the cost of a unit of electricity.

Solution:

(a) From equation 7.1. the annual repayment is £106.43.

The station generates 0.75 x 87609 = 6570 units.

The repayment cost is $\frac{106.43 \times 100}{6570}$ = 1.62 p/kWh

(b) From equation 7.1. the annual repayment is £176.19

The station generates 0.55 x 8760 = 4818 units.

The repayment cost is $\frac{176.19 \times 100}{4818}$ = 3.66 p/kWh

Fuel and other operating costs must be added to get the total generating costs.

It is obvious from this example that capital repayment costs can easily vary by a factor of two, or more, depending on the assumptions. If the original loan was made some years ago an appropriate calculation should be made each year to increase the annual repayment to allow for inflation. Costs occured in earlier years are known as 'historic' costs. If no allowance is made for inflation, the real value of the capital repaid and interest will never catch up with the outstanding balance.

The historic cost method was used in the UK until 1983. Before then, especially during the 1970s, official figures always showed an apparent substantial cost advantage for nuclear power. This is illustrated in Table 7.6., which gives an extract from the annual reports of the Central Electricity Generating Board for several different years.

Table 7.6

Generating costs for power stations commissioned after 1 April 1965. Relates to the 13 most recent coal-fired stations, two oil-fired stations and six Magnox nuclear stations[20,21].

Pence per kWh

Date	Magnox	Coal	Oil
1974-5	0.67	0.97	1.09
1978-9	1.02	1.29	1.31
1979-80	1.30	1.56	1.93
1980-81	1.65	1.85	2.62

These figures have been widely quoted from within the industry, in reports to government, and in some text books[22,23]. By 1980 a number of independent workers had started to examine the way in which these figures were produced. Jeffery[24,25] and Sweet[26] showed that if allowances were made for inflation and the standard treasury rates of interest, which were applied to other investment decisions, were used for calculating the costs of electricity, then a completely different picture would emerge. Magnox stations could be shown to be more expensive than coal and, in some cases, oil. This was confirmed by the Central Electricity Generating Board in 1983[19]. Table 7.7. shows new figures which were obtained using different sets of assumptions which are fully explained in the CEGB analysis[19]. The Opportunity Cost rate is broadly similiar to the annual interest rate r in equation 7.1.

Table 7.7

Generating Costs - Pence per kWh
1983 Cost Basis

Cost Basis	Magnox	Coal	Oil
Monetary Interest	2.06	2.05	3.22
Opportunity Cost 5%	3.37	2.28	3.87
Lifetime-to-date and Opportunity Cost 5%	2.60	2.19	2.49
Whole Lifetime and Opportunity Cost 5%	2.63	2.46	3.07

The CEGB pointed out that it is an open question whether any major emerging industrial technology would have become established had it been tested in its early stages against an opportunity cost of capital set at 5% in real terms.

When papers and reports produced twenty five years earlier are studied, they show that the industry only claimed that its first group of four Magnox reactors, Berkeley, Bradwell, Hunterston A and Hinkley Point A would produce electricity more cheaply than coal-fired power stations if the load factor were to be greater than 82%[27]. They hoped that 75% or better could be achieved. The full history of the UK nuclear programme is shown in Table 7.8., which also includes the world ranking figures and the load factors achieved by each reactor.

Table 7.9

Power station plant under construction at 31st March 1983

Power Station	Number of units and design net capability MWe	Year of start on site to probable completion of commissioning
Dungeness B	2 x 600	1966 - 84
Hartlepool	2 x 660	1968 - 83
Heysham 1	2 x 660	1970 - 83
Heysham 2	2 x 615	1981 - 88
Torness	2 x 700	1980 - 88

Data from Annual Report and Accounts 1982-83, Central Electricity Generating Board, London and South of Scotland Electricity Board, Glasgow.

Table 7.8

UK Commercial Nuclear Power Station Performance

Date of Regular Power Operation	Magnox Reactors	Installed Capacity MWe	Load factor and ranking: Cumulative	Load factor and ranking: Annual 1982/83
1962	Berkeley (CEGB), Gloucs.	320	70.3 (40)	18.6 (183)
1962	Bradwell (CEGB), Essex	347	61.7 (87)	50.0 (144)
1964	Hunterson A. (SSEB), Ayrshire	338	82.6 (5)	76.1 (56)
1965	Hinkley Point A (CEGB), Somerset	560	69.5 (46)	78.0 (53)
1965	Trawsfynydd (CEGB), Gwynedd	584.8	59.8 (99)	68.5 (84)
1966	Dungeness A (CEGB), Kent	571	56.8 (116=)	64.2 (99)
1966	Sizewell A (CEGB), Suffolk	652.5	60.3 (94)	45.6 (151)
1967/68	Oldbury A (CEGB), Gloucs.	626	55.4 (127=)	65.0 (95)
1971	Wylfa (CEGB), Anglesey	655	44.0 (168=)	63.5 (101)
	Advanced Gas Cooled Reactors			
1976/77	Hinkley Point B1 and B2, (CEGB), Somerset	2 x 660	42.9 (173)	60.9 (115)
1975/76	Hunterston B1 (SSEB), Ayrshire	660	41.0 (178)	71.1 (72)
	Hunterston B2 (SSEB), Ayrshire	660	33.2 (191=)	54.4 (135)

During 1970/71, all the Magnox Stations, other than Berkeley, were downrated

Data from Nuclear Engineering International, 28, 343, July 1983

The performance of the first group can now be compared with the original design plan to achieve a load factor of 75% or better. The weighted cumulative load factor for these four stations is 70.8%. In 1983 two were returning to full operation following a period when they were shut down for repairs[28]. The other two had annual figures of 76.1% and 78%.

The second major item in the calculation of the costs of nuclear electricity is known as the Works Costs. This includes fuel and transport costs and other costs such as handling, operation, repairs and maintenance. Most nuclear industry cost predictions have stated that nuclear fuel costs will be constant (in real cost terms) over a considerable period while the costs of the fossil fuels will continue to rise[17,26]. An examination of the statistics for the UK in 1983[29] shows that over the period from 1965 to 1983, which includes the major oil price rises in 1973-4 and 1979, nuclear fuel costs have risen very substantially compared with conventional fossil fuel costs. An extract from the full data is given in Table 7.10. The data for 1965/66 is given in historic figures.

Table 7.10

Works costs
Pence per kWh

	Nuclear		Conventional	
	Fuel including transport	Total costs	Fuel including transport	Total costs
1965/66	0.0650	0.1025	0.2233	0.2828
1982/83	1.395	1.75	1.948	2.3129

Data abstracted from Table 20 of reference 29.

In 1965/66 conventional fuel costs were 243% greater than nuclear fuel. In 1982/83 the figure was down to 40% greater. A similar trend is seen in total works costs. In 1965/66 conventional works costs were 176% greater than nuclear works costs. In 1982/83 the figure was 32% greater.

Experience in the United States from the early 1970s has shown that escalating capital costs combined with expensive breakdowns and the need to fit additional equipment has had a very serious impact on the industry. In November 1982, David Freeman[30], Managing Director of the Tennessee Valley Authority, pointed out that they had halted work on eight reactors over the past four years.

> ...because the cost of making them safe was simply going to be too much... The last three nuclear units TVA halted, were going to produce electricity estimated to cost about 13 cents per kilowatt-hour, more than triple our present average cost of about 4 cents per kilowatt-hour. ...in a sense TVA has not really completed any of its nuclear plants. At our Browns Ferry nuclear plant, which has been commercial for years, we still have more than 600 people at work making back fitting changes...

At the 1983 World Energy Conference in New Delhi it was reported that no new orders [for nuclear power stations] were expected in the United States before 1990[31].

The Indian experience is significant as it was the first nuclear programme in a developing country as well as being one of the most comprehensive. Hart[32] points out that their prior expectations of nuclear load factors in the order of 80% were unrealistic. This is confirmed by the history of their reactor performance up to 31st March 1983, which is given in Table 7.11 which also includes the world ranking figures.

Table 7.11

Indian Commercial Nuclear Power Stations Performance

Date of Regular Power Operation		Load factor Annual	Load factor Cummulative	Design Capacity MWe
1964	Tarapur 1	42.7 (156)	47.9 (159)	210
1969	Tarapur 2	37.2 (166	50.3 (152)	210
1978	Rapp 1	0 (190=)	34.0 (188=)	220
1981	Rapp 2	28.6 (176)	30.6 (195)	220

Data from Nuclear Engineering International[15].

Apart from Tarapur Boiling Water Reactors, which were constructed by the Americans and delayed by only one year, every Indian power reactor has been delayed for four to six years, with costs on most projects doubling[33]. Rapp and subsequent Indian reactors are Pressurized Heavy Water Reactors. Although assumptions favourable to nuclear power were used by Hart in calculating costs, the conventional power sources of coal and hydro were shown to be cheaper.

Radiation and Health

In principle, damage to health could be caused by radiation, perhaps through exposure to radioactive materials in minute quantities. The study of the effects of radiation is very complex. For example external radiation may occur from x-rays, gamma rays, alpha or beta radiation, neutrons, protons or high energy particles from outer space. The human body may be irradiated uniformly, the radiation may be highly concentrated or restricted to a limited area, as in medical applications. If radioactive material is inhaled, its location depends on its particle size, chemical nature and time in the body.

The unit of absorbed radiation dose is the gray (Gy) defined as an absorbed energy of one joule per kilogram. The former 'radiation absorbed dose' or rad is exactly one-hundredth of a gray. Different tissues have very different sensitivities to radiation. Doses of different types of radiation can be compared on the basis of the Dose Equivalent, obtained by multiplying the absorbed dose by a factor related to the stopping power of the radiation. The unit of Dose Equivalent is the sievert (Sv). The former unit, the rem, is exactly one-hundredth of a sievert. 5 millisieverts is the maximum permissible annual dose for the general public set by the International Commission on Radiological Protection (ICRP).

Natural background radiation levels vary according to the type of local geographical conditions and on the use of building materials. For example, in Aberdeen, where granite is used in building, the terrestial component of background radiation is 0.8mGy. In South-East England it is 0.4mGy. No changes in health have yet been attributed to differences in natural background radiation[34].

The ICRP points out that no radiation dose, however small, can be considered without risk, but the dose may be so small as to be negligible or unobservable. The internationally accepted dose limit recommended by the ICRP for personnel working in nuclear power stations is 50 millisieverts a year to the whole body. This will lead to risks comparable with those in industries considered 'safe'. The CEGB designs their nuclear plant to criteria intended to limit doses to one-fifth of this. The standards of the ICRP are endorsed in the UK by the National Radiological Protection Board and the Medical Research Council and form the basis of UK nuclear legislation and codes of practice. When the annual probably of death for males in England and Wales is studied, it can be seen that the increment of risk for a worker in the nuclear industry never exceeds a few tenths of one per cent of the risk due to natural causes[34]. Nine claims for compensation, two for leukaemia and seven for cancer have been made by workers at Sellafield since management responsibility was transferred to British Nuclear Fuels in 1971. By January 1984, compensation had been paid in six out-of-court settlements. Although liability was not admitted, the payment was made in recognition of the possibility of a cancer being linked to radiation[35].

Waste Disposal and the Management of Radioactive Waste

It was recognized from the earliest days of the nuclear industry that nuclear waste was very dangerous and that some radioactive materials could be potentially dangerous for thousands of years. In 1955 a UK Government White Paper (Cmd.9389) stated

> The disposal of radioactive waste products should not present a difficulty... The volume of waste will be small and great efforts are being made to determine the most economical methods of storing or disposing of it.

By the mid-1970s there as considerable public concern about leaks from the radioactive waste storage ranks at Hanford in the United States and at Windscale (now Sellafield) in the UK. The Sixth Report of the Royal Commission on Environmental Pollution concluded in September 1976 that

> ...there should be no commitment to a large programme of nuclear fission power until it has been demonstrated beyond reasonable doubt that a method exists to ensure the safe containment of long-lived highly radioactive waste for the indefinite future.

There is an important distinction between the different levels of radioactivity in nuclear waste. The industry considers three levels, high, intermediate and low[36]. Some low and medium level wastes, such as water from cooling ponds or gaseous effluents, are treated and discharged to the sea or to the atmosphere. Some fission products have a short half-life, only a few hours, and quickly decay into stable non-radioactive materials while still in the reactor. Some have half-lives of only a few months and are comparatively straightforward to deal with in the reprocessing plant as they become relatively harmless. Solid wastes can be buried on land or disposed of by sea dumping. By the time their containers corrode, the level of radioactivity is considered to be harmless. The discharge of these low level wastes is strictly controlled and is not considered to have harmed the environment. In the UK any discharges which result in an abnormal rise in the levels of radioactivity, such as the radioactive seaweed found near Sellafield towards the end of 1983, are the subject of strict investigations. High level nuclear waste is the most dangerous but exists in relatively small quantities. Two fission products, Strontium 90 and Caesium 137 have half-lives of 28 and 27 years respectively. A minimum of 600 years must elapse before their activity level is comparable with that of the original uranium ore. Plutonium 239 has a half-life of 24,400 years and needs to be isolated for over half a million years.

However it can only be harmful if taken in the body by breathing or swallowing.

British Nuclear Fuels Limited treats high level waste fission products in four stages

(i) The fuel is submerged in special water tanks, known as cooling ponds, for at least a year and, in some cases, up to ten years, to allow the intense radioactivity to die away.

(ii) After the cooling period the fuel is removed from the pond and can be mechanically cut or put into convenient lengths for reprocessing. In reprocessing, the uranium and plutonium are separated from each other and from the highly radioactive fission products. Plutonium is then stored for possible future use in the Fast Breeder Reactor. The fission products, which constitute between ½% and 3% of the spent fuel are concentrated for further storage as a liquid in special double-walled stainless steel tanks fitted with cooling coils.

(iii) When these products have cooled further, it is proposed that the wastes should be concentrated again and fused into a glass material in stainless steel containers and kept under water in a pond for about fifty years.

(iv) As the heat output would be considerably reduced at this point, the containers could be buried without artificial cooling until the remaining activity has decayed.

The first two stages have been used for a number of years. The French have manufactured a borosilicate glass in which wastes can be incorporated since 1979.

Any final containment position must be sufficiently stable and secure to allow the radioactivity to decay to a point where it poses no further hazards to the environment.

Waste disposal and management is clearly an area of considerable uncertainty. Nobody knows if it will be possible to manage the admittedly small volumes of highly active radioactive waste safely for several hundreds of years. One system of financial accounting, known as the discounted cash flow method, puts a negligible cost, in today's money, on activities which will be carried out in more than 100 years time. The fact that there will be nuclear waste to guard long after the energy power source has ceased to function, effectively creates a debt for future generations to repay.

Nuclear Power and Weapons Proliferation

From the 1950s the International Atomic Energy Agency was given two objectives - to promote the peaceful uses of nuclear energy and to provide safeguards against their misuse. At one time it was thought that reactor grade plutonium would not be suitable for nuclear weapons. This has been shown to be incorrect[37], although it would only be with extreme difficulty that a small and unreliable weapon could be prepared. In 1970, the Nuclear Non-Proliferation Treaty was drawn up. Since then it has been signed by at least 112 non-nuclear countries who have undertaken not to produce or acquire nuclear weapons. This is a major step towards the peaceful uses of nuclear energy. But some 50 nations have refused to sign the treaty. Of these, it is believed that Israel, South Africa, Pakistan and Argentina may be in the process of acquiring a nuclear weapons capability.

The International Atomic Energy Agency has been regarded as ineffectual because it works on the assumption that the peaceful uses of nuclear energy can be promoted without military technology also being linked to it. A major weakness of the Nuclear Non-Proliferation Treaty is that countries who do not possess nuclear weapons will ask why it is unreasonable to expect them not to develop nuclear weapons if the major nuclear countries intend to keep their monopoly. Even if a major nuclear country, such as the United Kingdom, were to close down its nuclear power programme it is difficult to see how this would prevent or discourage a less developed country from building its own nuclear weapons.

The growth of nuclear weapons held nationally can only increase the threat of their use.

Civil Liberties

Issues relating to the security demands of the nuclear power industry cannot be divorced from the general security situation facing the world in 1980s. Restriction on certain civil liberties are accepted in many countries as part of the price that must be paid for safeguards against international terrorism. In the United Kingdom workers in the nuclear power industry are subject to detailed screening on appointment. They are also directly subject to the Official Secrets Act, but so are many thousands of government employees in a wide variety of occupations. It is also not widely appreciated that every citizen in the UK is also subject to the Official Secrets Act.

The Atomic Energy Authority (Special Constables) Act 1976 has given the UKAEA a Special Constabulary. Their duties include the guarding of plutonium and its related installations. They have powers to arrest and search suspected terrorists. The existence of this force has been considered by some people to be deeply disturbing on the grounds of civil liberties. Others welcome the security from possible terrorist activities which such a force provides.

The Methodology of Risk Assessment

Methods of risk assessment, such as the 'event tree' and 'fault tree' methods, attempt to identify the course of all possible accident sequences, and then give estimates of the individual probability of failure. The best known assessment of this type was carried out by Professor Norman C. Rasmussen of the Massachusetts Institute of Technology in a three year study sponsored by the Atomic Energy Commission in the United States[38]. The study involved over 60 investigators and consultants and about 70 man-years of effort. The probability of a core meltdown was originally estimated to be about one in ten thousand per reactor year, with the chance of a radiation release resulting from a meltdown being 10^{-5}, giving the overall chance of a catastrophic acident as 10^{-9} per reactor year, or one chance in a million if there were a thousand reactors in operation. To place this risk in context, a table giving the chances of a fatal accident to an individual in the United States in the year 1969 was also published. Extracts from this table are shown in Table 7.12.

While these figures look very reassuring, Lord Ashby of Brandon[39] points out that

> ...probability theory applied to human behaviour is invalid if groups of individuals can influence the response...

When Blackith[40] applied the Rasmussen techniques to some accidents which had already happened, he found one accident which should only occur in a million years and another which had a probability of twice in a million million million years. In January 1979 the United States Nuclear Regulatory Commission had stated that it could no longer 'regard as reliable' its own safety study[5].

Table 7.12

Chances of a fatal accident, USA, 1969

Accident type	Total Number	Individual chance per Year
Motor Vehicle	55,791	1 in 4,000
Falls	17,827	1 in 10,000
Fires and hot substances	7,451	1 in 25,000
Drowning	6,181	1 in 30,000
Air Travel	1,778	1 in 100,000
Electrocution	1,148	1 in 160,000
Tornadoes/Hurricanes	184	1 in 2,500,000
All accidents	111,992	1 in 1,600
Nuclear reactor accidents (100 plants)	0	1 in 5,000,000,000

Safety Features in Design and Operation

The first fact to appreciate is that a nuclear reactor cannot explode like an atom bomb. The content of the fissile material in the core is far too low and diffuse. An independent safeguard in the UK is the Nuclear Installations Inspectorate. This was established in 1960 to assess the safety aspects of nuclear plant designs prior to approval for construction and operation and to carry out inspections at sites during the construction and operating stages. The primary task of the NII is application of the relevant statutory provisions of the Nuclear Installations Act 1965 and the more general provisions of the Health and Safety at Work Act 1974 to nuclear power stations and sites used for the manufacture and reprocessing of nuclear fuel and the storage of radioactive waste[41]. The 1974 Act allows any one of the nuclear inspectors to stop a nuclear power station by revoking its operating licence. In 1975 the NII was transfered to the Health and Safety Executive (HSE).

Concern was expressed about the problems of recruiting additional staff to deal with all the tasks facing the Inspectorate in 1978[42]. In 1980 a newspaper report[43] drew attention to the difficulties arising from the ageing of existing plants and the need to evaluate the complex problems of the PWR. The government response was that 'shortages of inspectors would not mean a slowdown in work on nuclear power programmes'.

In designing nuclear reactor plant to meet the criteria arising from the UK fault analysis techniques, a design philosophy usually described as 'Defence in Depth' has been adopted. As far as possible, mechanisms are selected with inherent characteristics and properties which provide self-limiting safety characteristics. Four main guiding principles have been adopted in the UK[44].

(a) Diversity
Diversity is achieving a given objective in different ways. It could mean using two different sensing instruments for the same function. For example the presence of the coolant in the reactor core could be detected by its flow rate or its pressure. Alternative equipment operating by entirely different principles could also be installed.

(b) Duplicating or trebling equipment
This acts as an insurance against the failure of any single important element to operate or to operate correctly. It is common practice to install three separate pumps

of equal capacity where functionally only one is required to do the job. Even if one pump is undergoing maintenance and another fails to work, there is still a third available to carry out the required function. Similarly, extra lines of instrumentation are installed to guard against the possible incorrect operation of a single line.

(c) Physical segregation
The physical segregation of equipment and systems is adopted to limit the extent of failures caused by localised floods or fires to one of the available lines of instrumentation.

(d) Fail safe design
This has the objective of ensuring that the failure of any piece of equipment should always result in a safe situation, both in respect of that particular item of equipment and with no attendant safety implications for the plant as a whole. For example, control rods should drop down under gravity into the reactor core if the power supply fails, thus shutting it off safely.

While the UK industry has always stated publicly that both human failures and plant troubles are bound to occur, it believes that the design approach is such that there are always sufficient protecting sequences to contain any incident and limit its effect to an acceptable level. Recently a new principle was mentioned by the Chairman of the CEGB, Sir Walter Marshall[45]. He stated that all UK reactor operators have been instructed to think for 30 minutes in the event of an emergency situation so that they can fully understand what has happened before taking any action. Marshall went on to comment that the accident at Three Mile Island, described in the following section, could not have occured in the UK because of the superior UK design and operating philosophy.

The Three Mile Island Accident

Three Mile Island Reactor, Unit 2, is a PWR with a rated capacity of 956 MWe. On the 28th March 1979 a sequence of events occured which resulted in the worst accident, in financial terms, in the history of the nuclear industry. The full sequence of events leading to the accident has been fully recorded[46]. In reviewing them, it appears that there were some unresolved problems before the accident sequence commenced. There was a leakage of reactor coolant which was exceeding the specified limits, but the operators did not appreciate this. About two days before the accident two valves in the auxiliary feedwater system were shut down during routine servicing and not reopened. This contravened the regulations. During the eleven hours prior to the accident the operators had been trying to clear a blockage in a pipe. In trying to clear the blockage they caused a condensate pump to stop working. This immediately made the main feedwater pumps stop, as designed, causing a failure in the main feedwater system. All three of the auxiliary feedwater pumps then started in accordance with normal procedures, but because the two valves had been incorrectly left shut, there was no flow. The operators were not immediately aware of this because their view of the valve position indicators in the control room was blocked by some poorly positioned labels. Subsequently further operating mistakes were made. It was not appreciated that a relief valve had stuck in the open position and that this was causing a serious loss of coolant. Eventually the core was exposed. This resulted in a leak of radioiactivity, first into the auxiliary building and finally into the outside environment. The severly damaged reactor core was surrounded with highly dangerous radioactive debris. This had not been cleared by the end of 1983.

The main conclusion of the President's Commission[47] set up to investigate the accident was as follows:

> To prevent nuclear accidents as serious as Three Mile Island, fundamental changes will be necessary in the organization, procedures and practices - and above all - in the attitudes of the Nuclear Regulatory Commission and, to the extent that the institutions we investigated are typical, the nuclear industry.

David Freeman[9], Managing Director of the Tennessee Valley Authority, looking back at the event some four years later commented

> ...we were told it could not happen. It's not good enough to say no one was injured. The industry said you could not loose the cooling water and that reactor cooked, part of elements melted, it did self-destruct...

The nuclear industry can state quite correctly that despite a very unlikely chain of accidents, computed at less than one chance in 10 million for each year of reactor operation, no major catastrophe occurred.

Research and Development with Fast Breeder Reactors

As early as 1946 it was appreciated that a different type of reactor could be developed, 'the object of which is to increase the amount of fissile material available'[48]. All the plutonium-239 (a fissile material) in use in the civil nuclear cycle has been produced through the operation of thermal reactors. It is created when uranium-238 absorbs a neutron. The term Fast Breeder Reactor (FBR) is used to describe the class of reactor which uses fast neutrons with a uranium based fuel which contains up to 30% of plutonium-239, and which is capable of breeding new fissile material. The core of a FBR contains a high concentration of fissile material. This sustains the chain reaction without a moderator. Liquid sodium, which has very good heat transfer properties at high temperatures, is used as the coolant in the primary circuit. This radioactive sodium transfers heat to a second sodium circuit which, in turn, transfers heat to a conventional steam plant. While sodium is well-known to react fiercely with both air and water, it is operating at very moderate pressures in the FBR, just slightly above atmospheric, and with a sensible design philosophy should be quite safe.

One method likely to be used for starting up a FBR, is to use some of the plutonium accumulated from the operation of thermal reactors. Once the reactor has started it is possible to create new plutonium by surrounding the core of the FBR by a 'blanket' of non-fissile uranium-238, which is also left over as spent fuel from the thermal reactors. The conversion of uranium-238 to plutonium is an optional process. Without it, a 1000 MWe reactor would consume about 230 kg of fissile plutonium-239 in a year. With the blanket in operation it could produce up to 190 kg per annum[48].

The time taken for a FBR to breed enough new plutonium to fuel a second reactor and the ancillary fuel cycle is known as the doubling time of the reactor. In the early days of the research programmes it was thought this might take about ten years, but as Marshall[48] points out, fast breeder reactors do not breed fast, they simply use fast neutrons and breed rather slowly. Various estimates of doubling times suggest that thirty years was a reasonable figure to assume in the early 1980s.

Three countries, the Soviet Union, France and the UK, have experimental fast breeder reactors with outputs greater than 250 MWe. The nearest to a commercial system is the 1200 MWe French Super-Phenix at Creys-Malville, due to commence operation during 1984. In the UK, the Prototype Fast Reactor (PFR) of 270 MWe gross, at Dounreay, first generated electricity in 1976. Its cummulative load factor of 10.6%, achieved by the 31st March 1983, should not yet be compared with normal commercial reactors because of the needs of the experimental programme. At the fuel side of the

FBR substantial achievements were reported in 1983[15]. Spent fuel can be recycled from its removal from the reactor, through the reprocessing and fabrication stages, in under a year. This will reduce the projected cost of operation as the original time was estimated as up to three years.

A fast reactor programme would, in theory, be able to make optimum use of the energy potentially available in natural uranium. About 50 times more energy is theoretically available from up-grading uranium-238 in the FBR cycle, compared with the original energy obtained from using natural uranium in thermal reactors. In the UK there were 20000 tonnes of depleted uranium in store by 1983, which was claimed by the UKAEA[49] to have an energy potential comparable with the country's economically recoverable coal reserves or five times the oil likely to be recovered from the North Sea. If doubling times in practice became much longer than 30 years, a programme of replacing thermal reactors with FBRs would have a limited value. The reactor's useful life would be over before sufficient fuel for the next reactor had been created and some of the theoretically available energy in the depleted uranium stocks could never be used.

Research and Development with Fusion Reactions

A fusion reaction is one in which the light nuclei of two atoms combine to form a heavier nucleus with a release of energy. Fusion reactions occur naturally in the sun and the stars. There are many possible fusion reactions. The most promising being considered for practical fusion power uses the heavy isotopes of hydrogen, deuterium and tritium[50]. Deuterium is a common stable isotope which occurs naturally in seawater. Tritium does not occur naturally and uses lithium as its primary fuel. Identified world reserves of lithium are in the order of 7 million tonnes, compared with present usage of some five thousand tonnes per annum, mainly for lubricants and ceramics[51]. If the two isotopes were to be consumed in equal quantities, these reserves are equivalent to at least a thousand times today's total world annual energy consumption. Further immense reserves of lithium exist in the oceans, perhaps extending the reserves by a factor of 10,000[52]. This is the reason for fusion research.

Fusion reactions are very difficult to achieve because the electric charges of the nuclei cause a mutual repulsion. The repelling force can be overcome if the nuclei approach one another at very high velocities. One way of achieving this is to heat a gaseous mixture of deuterium and tritium to temperatures approaching 10^8K. At high temperatures the negatively charged electrons in a gas effectively reach escape velocity and fly off. The hot gas in transformed to an assembly of charged particles, an ionized gas or plasma which is influenced by electric and magnetic fields. Other plasma conditions must also be satisfied for the rate of nuclear fusion energy generation to be greater than the total energy input. These are that the plasma must be confined for a certain minimum period, about 0.5 to 1 seconds for the deuterium-tritium reaction, and that the plasma density in nuclei per m^3 should be such that the product of (minimum time period) x (plasma density) is greater than 10^{20}[53]. At room temperature this ion density corresponds to a pressure of 3×10^{-3} Torr, a good vacuum pressure. There are three basic steps which must be taken to achieve electricity generation from controlled nuclear fusion[50].

(i) To create and heat plasma to temperatures in excess of 10^8K.
(ii) To hold enough plasma away from container walls for long enough to allow abundant reactions to occur.
(iii) To design a practical fusion reactor to produce electricity economically.

By 1983 work was concerned only with the first two steps. The most promising approach for confining a plasma is by the use of a magnetic field. Most systems currently use toroidal, or ring-shaped, confinement systems, with the most highly developed known as 'tokamaks'. A second approach uses the inertia of the fuel to hold it together while the reactions take place. The use of lasers is being studied in a number of laboratories. In December 1983 it was reported[54] that workers at the Massachusetts Institute of Technology had achieved the 'break even' point, where they were generating as much energy as was used in heating the gas. They fired pellets of solid deuterium at a temperature of about -270°C into the plasma, which was at too low a density for the fusion reaction. The deuterium vaporized and the reaction occurred.

Another advantage of fusion energy systems is that neither the primary fuels nor the final reaction products are radioactive. However, the intermediate fuel, tritium, has a half-life of 12.6 years and the high energy neutrons released from the fusion reaction will make the structure of the reactor radioactive[50]. The quantity of active material in a fusion reactor would be up to 100 times less than in a fully fuelled fission reactor. Any loss of control of the reaction would not be dangerous as it is self-limiting. Fusion reactors would not generate plutonium directly although it is theoretically possible[51].

Extensive international collaboration in fusion research dates from the period 1956-60. The research work was declassified in 1958 and freed for international collaboration[55]. By 1983 four large tokamaks were being commissioned or nearing completion -TFTR (USA), the Joint European Torus JET based at Culham, Oxford (Western Europe), JT60 (Japan) and T15 (USSR). TFTR commenced operating in December 1982 followed by JET in mid-1983. TFTR and JET are designed to generate several megawatts of heat for a few seconds in the latter part of their programme.

The leading research groups emphasize that the development of controlled nuclear fusion is still in the research stage, but that experience so far has been encouraging[55,56].

Uranium Reserves

Estimates of the reserves of Uranium are at concentrations which would be economic to mine vary according to the assumptions of mining costs. The amount of uranium oxide which could be produced ranges from 2.5 million tonnes to about 8 million tonnes. This could provide at least ten times the annual primary energy equivalent of the total world electrical output in 1983.

Since 1970 uranium production has been greatly in excess of consumption. For most of the period up to 1983 the scale of over production has been at least 50%. The most significant contributory factor seems to have been the continued tendency towards over-estimation in forecasts of nuclear capacity. Even as recently as 1979, the forecasts for 1990 are now unlikely to be reached until 2000[15]. By 1980 annual production was twice consumption. By 1983 consumers were believed to have about four years forward consumption in stock. The world price of uranium oxide had slumped to about half its 1978 levels in five years.

Nevertheless, in the longer term economically available supplies of uranium ore are limited, and the total maximum generation capacity of thermal reactors is unlikely to exceed 1300 GWe[57].

Summary

Up to 1970 the growth of the nuclear industry was rapid. Since then a number of factors have combined to slow down this growth. Among these factors can be listed the reduced forecasts of growth in the demand for electricity, financial constraints, lack of confidence in performance, uncertainty about possible future regulations and a major reassessment of the economics compared with coal-fired stations[15]. The promise held out for the future in both the Fast Breeder and Fusion reactors has resulted in very substantial research funding. This has meant that funding for other forms of energy research or conservation measures has been limited. There are certain social costs to accept with any nuclear programme. This is a subjective issue. The main justification for a nuclear programme in the future could be on the grounds of long-term security of supply. Until the end of the 20th century nuclear power can only make a relatively minor contribution to the total world demand for energy.

References

1. Soddy, Frederick, Matter and Energy, Williams and Norgate, London, 1912.

2. Soddy, Frederick, Cartesian Economics, Two lectures delivered in London University, November 1921; Henderson, London, 1924.

3. Tombs, Francis, A Review of Nuclear Power in the United Kingdom, Lecture, The Electricity Council, London, 1977.

4. Patterson, Walter C., Nuclear Power, Penguin Books Ltd., 1976.

5. Time, 18, 9 April, 1978.

6. Theoretical Possibilities and Consequences of Major Accidents in Large Nuclear Power Plants, WASH-740, U.S. Congressional Joint Committee, March 1957.

7. B.P. Statistical Reviews of (a) World Energy 1982 and (b) the World Oil Industry 1978. The British Petroleum Company plc, London, 1983 and 1979.

8. Guide to N.P.C. Report, U.S. Energy Outlook, National Petroleum Council, Washington, 1972.

9. Sizewell Under Pressure, Horizon, BBC T.V. programme, 10th January 1983.

10. Ehrich, T., in The Wall Street Journal, 3rd May 1973.

11. Lovins, Amory B., Is nuclear power necessary?, Friends of the Earth, London, 1979.

12. Foley, Gerald, The Energy Question, Second Edition, Penguin Books, 1981.

13. Blevins, R.D., Flow-induced vibration in nuclear reactors: a review. Progress in Nuclear Energy, 4, 25-49, 1979.

14. Wicker, T., in the International Herald Tribune, 3-4 March 1979.

15. Nuclear Engineering International, 28, 343, July 1983.

16. Buhl, Harold, R. Creative Engineering Design, Iowa State University Press, 1960.

17. Sweet, Colin. The PWR and Energy Policy. Issues in the Sizewell 'B' Inquiry, Centre for Energy Studies, Polytechnic of the South Bank, London, 1983.

18. Annual Report and Accounts 1982-83, South of Scotland Electricity Board, Glasgow, 1983.

19. Analysis of generating costs, Central Electricity Generating Board, February 1983.

20. Annual Report and Accounts, 1978-79, 12, CEGB, Sudbury House, 15 Newgate Street, London, EC1A 7AU, July 1979.

21. Costs of producing electricity, Central Electricity Generating Board, July 1981.

22. Lewis, John L., Nuclear reactions and nuclear power, in Energy, Book F, Science in Society, Heinemann Educational Books Ltd., London and the Association for Science Education, 1981.

23. Blair, Ian M., Taming the Atom, Adam Hilger Ltd., Bristol, 1983.

24. Jeffery, J.W., The real costs of nuclear power in the UK, Energy Policy, 8, 4, 344-6, 1980.

25. Jeffery, J.W., The real cost of nuclear electricity in the UK, Energy Policy, 10, 2, 76-100, 1982.

26. Sweet, Colin. The price of nuclear power, Heinmann Educational Books, London 1983.

27. Fletcher, P.T., 'Nuclear Power', Proc. Conf. Economic Aspects of Fuel and Power in British Industry, (5-7 November 1958), 124-147, Manchester University Press, 1960.

28. Varley, J., Meeting the challenge of the Magnox weld defects, Nuclear Engineering International, 288, 339, April 1983.

29. Handbook of Electricity Supply Statistics, The Electricity Council, London, 1983.

30. Freeman, David, quoted in The Case for CANDU, Paper VB01 (revised), presented at the Sizewell Inquiry, Lord Bowden of Chesterfield, May 1983.

31. Nuclear club seeks third world members. Electrical Review, 213, 13, 16-17, 1983.

32. Hart, David. Nuclear Power in India. A comparative analysis, George Allen and Unwin, London 1983.

33. Poneman, Daniel. Nuclear Power in the Developing World, 55-6, George Allen and Unwin, London, 1982.

34. Mayneord, W.V. and Wheatley, B.M., Nuclear Power: putting the risks in perspective, CEGB Research, London, January 1981.

35. News item, The Daily Telegraph, 4th January 1984.

36. Disposal of Radioactive Waste in the Deep Ocean, British Nuclear Forum, London, 1983.

37. Rotblat, J. Nuclear Power and Weapons Proliferation, Issues in the Sizewell 'B' Inquiry, Centre for Energy Studies, Polytechnic of the South Bank, London 1983.

38. Reactor Safety Study, WASH-1400, Nuclear Regulatory Commission, Washington, October 1975.

39. Lord Ashby of Brandon, Introductory Address, Conf. Energy and our Future Environment, Institute of Energy, London, November 1983.

40. Blackith, R.E., in A Nuclear Ireland, Irish Transport and General Workers Union, Dublin 1978.

41. Nuclear Energy in the UK, Fact Sheet 5, Department of Energy, March 1977.

42. Health and Safety, Nuclear Establishments 1977-78, Health and Safety Executive Report, 1978.

43. Article, The Guardian, 14th February, 1980.

44. Nuclear Safety, British Nuclear Forum, London 1978.

45. Marshall, W., Nuclear Energy and the Environment, Conf. Energy and our Future Environment, Institute of Energy, London, November 1983.

46. Rubinstein, E. Three Mile Island: The accident that should not have happened, IEEE Spectrum, 16, 11, 1979.

47. Kemeny, J.G. (Chairman), Report of the President's Commission on the Accident at Three Mile Island, Pergamon Press, New York 1980.

48. Marshall, W. Fast Reactors. Atom, 287, September 1980.

49. Fast Reactors: potential for power, UKAEA, London, 1983.

50. Nuclear Fusion, UKAEA, September 1983.

51. Hancox, R. Fusion-Power for the 21st Century, Culham Laboratory, Abingdon, Oxford 1979.

52. Gibson, A. The JET project. Atom, 254, December 1977.

53. Hunt, R.R. et al. Engineering aspects of fusion experiments, Chartered Mechanical Engineer, 26, 3, 55-60, 1979.

54. Osman, T. Scoring shots in fusion game, The Sunday Times, 11th December 1983.

55. Pease, R.A., International Fusion Research, Atom 315, January, 1983.

56. Fusion: The European Scene, Atom, 308, June 1982.

57. Rotblat, J. The Guardian, 29th May 1980.

Further Reading

In the UK the British Nuclear Forum, 1 St. Alban's Street, London, can provide a wide variety of up-to-date technical literature on all aspects of nuclear power and can also advise on specialist nuclear information sources. The monthly magazine Atom, produced by the UKAEA, often contains excellent summaries of current research and development programmes. A concise history, the Development of Atomic Energy, Chronology of Events 1939-1978, was published by the Authority Historian's office of the UKAEA in 1979. Dr. I.M. Blair's Taming the Atom, Adam Hilger Ltd., Bristol, 1983 gives a non-technical personal account of the development and possible future of nuclear power. Blair works with the Nuclear Environment Branch at Harwell.

Professor S.E. Hunt's Fission, Fusion and the Energy Crisis, Pergamon Press, 1980, covers the main technical aspects in some depths while the Royal Institution's Forum, Nuclear or Not?, Heinemann, 1978, shows well balanced arguments for and against nuclear power.

Aspects of safety are covered objectively by Sir Alan Cottrell in How Safe is Nuclear Energy, Heinemann, 1981, a book written particularly for the layman. Dr. David Collingridge's Technology in the Policy Process, Frances Pinter (Publishers), 1983, examines the social and political control of nuclear power while Colin Sweet's The Price of Nuclear Power, Heinemann, 1983, makes a very forceful economic case for the immediate abandonment of the entire nuclear programme. Sweet is Director of the Centre for Energy Studies at the Polytechnic of the South Bank, London, where a number of economic, social and political assessments of nuclear power have been published. The booklet Analysis of Generating Costs, CEGB, 15 Newgate Street, London, February 1983 explains their Net Effective Cost method, which includes systems savings as well as the costs used in the analysis of generation costs. This method is used to support their claim that the proposed Sizewell 'B' PWR stands out as the best choice for large-scale electricity generation plant. Among earlier works, Walter Patterson's Nuclear Power, Penguin Books, 1976, is critical of the industry, while Sir Fred Hoyle's Energy or Extinction, Heinemann, 1977, is strongly in favour.

Exercises

1 By the end of 1982 world nuclear energy consumption was the equivalent of 216.6 million tonnes of oil and increasing at a growth rate of 10% per annum. World water power consumption was the equivalent of 446 million tonnes of oil and increasing at growth rate of 3.5% per annum. When would the nuclear and water power consumption rates be equal if these growth rates ramained constant? What would the annual consumption rate of both resources be then?

2. The half-life of Strontium 90 is 28.1 years. How long will it take for its level of radioactivity to fall to one tenth of its original value?

3. The total capital charges for a new 1200 MW nuclear reactor, including provision for decommissioning, are £1,860 million. Determine the cost of electricity generation in pence per Kilowatt hour with the following set of assumptions:

   (i) A 5% rate of interest is used over a lifetime of 30 years to repay the capital charges.
   (ii) Fuel and operating costs remain constant, in real terms, at 1.05 p/kWh.
   (iii) The cummulative load factor is 67%.

4. Low level solid radioactive wastes have been disposed at sea by the UK since 1949. Action by some trades unions prevented this during the early 1980s. Examine the latest position in the UK and throughout the world relating to the disposal of low level solid radioactive wastes at sea.

5. During a discussion on the future of nuclear power in the UK held in 1958, a speaker commented[27]

   > ...what really struck me was the enormous figures that you gave for research and development ...a figure ten or twelve times as great as that put into research and development in all the three conventional nationalized fuel industries ...who decides this allocation of the scarce resources of our capital and skilled labour, and is this the right allocation?

An extract from the reply was as follows:

> ...we have to indulge in a very considerable act of faith in the future. We have to work this out on a speculative basis ...There have been discussions between power authorities and the AEA and the Generating Boards, and economists have looked at this. We, like the other industries, try to sell our wares to a degree, and I think you can give us that privilege, and out of this the pattern has emerged.

(a) To what extent can it be said today that the 'very considerable act of faith in the future' was justified?

(b) Has there been any significant change in the attitudes of the industry, as expressed in this extract from the reply, during the past 25 years?

(c) What is the annual figure for all nuclear research and development work in the UK this year? How does this compare with that spent on all other energy research and development work in the energy industries and on energy conservation measures?

6. The development of the nuclear industry in France is regarded by some analysts as a barometer for the future of the industry. A report published in 1983 stated there was no real need for any nuclear plant orders before 1987 at the earliest, and before 1991 if the lowest forecasts of demand were correct. French industry claimed it must have at least three nuclear orders a year if it were to remain active in research and in the export market. Examine the developments in the French nuclear industry since 1983. How has the Super-Phénix at Creys-Malville performed? What influence might this have on the future of the Fast Breeder Reactor?

7. What is radiation? What percentage of the total radiation to which the population of the UK are exposed consists of natural background radiation? What is the percentage due to the nuclear power industry?

8. The inquiry into plans to build a PWR at Sizewell in the UK commenced in January 1983. By January 1984 the CEGB had presented its case and many opposition witnesses had still to be heard. What are the main arguments in favour of the PWR and the most significant arguments against it? To what extent will the decision influence the future of nuclear power generation in the UK?

Answers

1. In 11.58 years ie. 1994; 670.6 m.t.o.e.
2. 93.33 years
3. 2.77 p/kWh
7. About 66%, Less than 0.2%

8. SOLAR ENERGY

Introduction

The importance of solar energy as a major energy source has been recognized for many years[1], but it is only in the past decade that substantial research, development and demonstration projects have been pursued throughout the world. Some of the major topics are described in this chapter with a brief outline of projects chosen to illustrate a few of the applications. The need for a new approach to the provision of energy was summarised by the author in an address at the 1975 meeting of the International Solar Energy Society at Los Angeles.

> It is now generally acepted that the exponential growth of energy consumption which has been experienced for many years cannot continue indefinitely. Solar energy is by far the most attractive alternative energy source available for the forseeable future. Quite apart from its non-polluting qualities, the amount of energy which is available for conversion is several orders of magnitude greater than all present world requirements....

The two major problems are the relatively low intensity of solar radiation at the earth's surface, about 1 kWm^{-2} is the maximum within an hour or two of mid-day on clear days, and its intermittency. An appreciation of the way in which the levels of solar radiation can vary in different geographical locations is essential for any assessment of a solar application. This can be studied by two methods. The first involves measurements from a radiation monitoring network and the second is based on the use of physical formulae and constants.

Solar Radiation

Direct solar radiation, I, is the solar radiation flux associated with the direct solar beam from the direction of the sun's disc, which may be assumed to be a point source, and is measured normal to the beam (that is on a plane which is perpendicular to the direction of the sun). Diffuse radiation, D, reaches the ground from the rest of the whole sky hemisphere from which it has been scattered in passing through the atmosphere. Global solar radiation, G, includes all the radiation, direct and diffuse, incident on a horizontal plane. The distribution of diffuse radiation is not uniform over the whole sky hemisphere and is more intense from a zone of about 5° radius

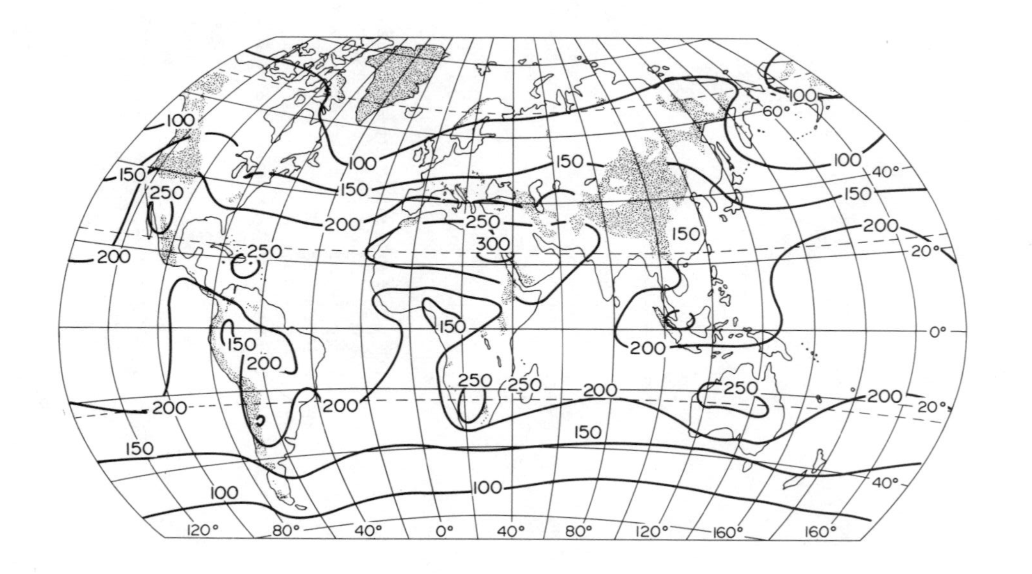

Figure 8.1 Annual mean global solar radiation on a horizontal plane at the earth's surface, Wm^{-2} averaged over 24 hours (based on [43] by kind permission of UK-ISES)

surrounding the sun. This is known as circumsolar radiation. Radiation may also be reflected from the ground onto any inclined surface, though this is very difficult to assess. The relationship between direct radiation, I, the diffuse radiation, D, and the global radiation, G, is given by

$$G = D + I \sin \gamma \tag{8.1}$$

where γ is the solar altitude above the horizon.

Solar radiation, or irradience, received on the earth's surface averaged over 24 hours on each day throughout the year is shown in Figure 8.1. The average in the UK is close to 100 Wm^{-2} while in the Arab countries, such as Egypt and the Sudan it is close to 300 Wm^{-2}. A detailed look at the mean daily totals of solar radiation, averaged for each month, in London and Khartoum is shown in Figure 8.2.

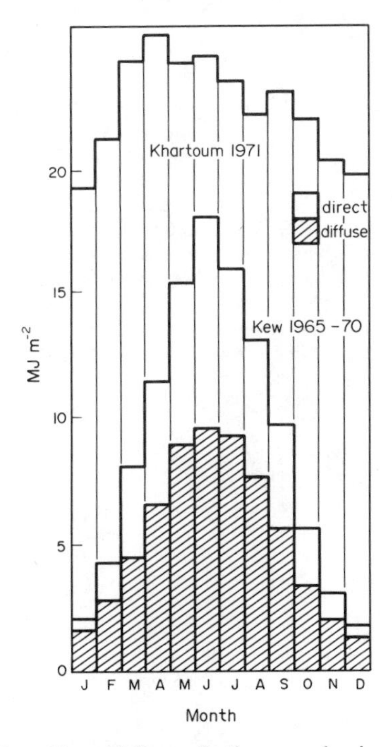

Figure 8.2 Mean daily radiation on a horizontal plane

In the UK and other Northern European countries there is about a four-fold variation between the worst three-month winter period and the best three-month summer period. About half the radiation in Northern Europe is diffuse. In many of the Arab countries there is only a small variation from one month to the next and there is a greater proportion of direct radiation. The worst month in the Sudan has a considerably greater total than the best month in the UK. Many applications, for example in architectural design and housing, require a knowledge of the total radiation on an inclined surface facing in any direction, while the only available data is the total global radiation on a horizontal surface in the same location or within a reasonable distance. This can be obtained by successive calculations from a complete solar radiation data base or from suitable radiation models. One example of the model approach was shown by Sayigh[2] who used well-known climatological data to develop an empirical formula from which an annual total radiation curve may be calculated. The formula includes factors for latitude, altitude, proximity to sea or lakes, the ratio of actual sunshine hours to the length of the day, mean relative humidity and monthly mean maximum temperature. This method can be applied to any location, but should be checked with real data where possible.

An alternative approach is to select an 'example year' for the particular country, using existing weather data. A number of organizations in the UK, including the Electricity Council, the National Coal Board, the British Gas Corporation and the Building Services Research and Information Association, have recommended that the year October 1964 - September 1965 should be selected as an example year for comparative energy demand calculations. This proposal has been described in detail by Holmes and Hitchin[3].

Flat Plate Collectors

Solar energy can be easily converted into useful heat which could provide a significant proportion of the domestic hot water and space heating demand in many countries. One of the drawbacks in high latitude countries, such as the UK, is that there are many days in the winter months when the total radiation received will be too small to make any useful contribution. The most widely known and understood method for converting solar energy into heat is by the use of a flat plate collector for heating water, air or some other fluid. The term 'flat plate' is slightly misleading and is used to describe a variety of different collectors which have combinations of flat, grooved and corrugated shapes as the absorbing surface, as well as various methods for transferring the absorbed solar radiation from the surface of the collector to the heated fluid. Many different types of collector have been built and tested by independent investigators over the past ninety years, the early work being carried out mainly in the USA, Australia, Israel, the UK and South Africa[1]. Tests were carried out in specific locations with wide variations in test procedures and in the availability of solar radiation. The main objective of these tests has been to convert as much solar radiation as possible into heat, at the highest attainable temperature, for the lowest possible investment in materials and labour[4].

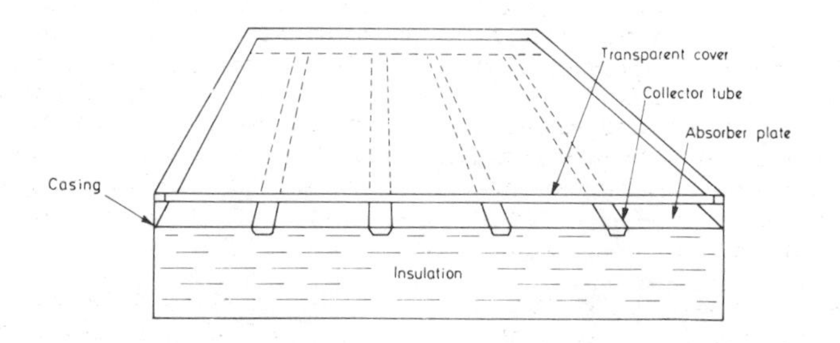

Figure 8.3

The majority of flat plat collectors have five main components, as shown in Figure 8.3. These are as follows:

(i) A transparent cover which may be one or more sheets of glass or a radiation-transmitting plastic film or sheet.
(ii) Tubes, fins, passages or channels integral with the collector absorber plate or connected to it, which carry the water, air or other fluid.
(iii) The absorber plate, normally metallic and with a black surface, although a wide variety of other materials can be used, particularly with air heaters.
(iv) Insulation, which sould be provided at the back and sides to minimize the heat losses.
(v) The casing or container which encloses the other components and protects them from the weather.

Components (i) and (iv) may be omitted for low temperature rise applications, such as the heating of swimming pools.

Flat plate collectors can also be classified into three groups according to their main applications as follows:

(i) Applications with a very small rise in temperature, such as in swimming pools where the collector needs no cover or insulation at the back or sides. A high rate of flow is maintained to limit the temperature rise to less than 2°C.
(ii) Domestic heating and other applications where the maximum temperature required is not more than 60°C. Insulation at the back and at least one transparent cover are necessary.
(iii) Applications such as process heating or the provision of small scale power, where temperatures considerably above 60°C are necessary. A more sophisticated design approach is needed to reduce heat losses from the collector to the surroundings.

One basic equation, the Hottel-Whillier-Bliss equation[5] expresses the useful heat collected, Q, per unit area, in terms of two operating variables, the incident solar radiation normal to the collector plate, G_c, and the temperature difference between the mean temperature of the heat removal fluid in the collector, T_m, and the surrounding air temperature, T_a, as follows:

$$Q = F[(\tau\alpha)G_c - U(T_m - T_a)] \tag{8.2}$$

where F is a factor related to the design of the collector plate and the effectiveness of heat transfer from the collector plate to the heat removal fluid.

The transmittance-absorptance product $(\tau\alpha)$ takes account of the complex interaction of optical properties in the solar radiation wavelengths[5]. It is actually some 5% greater than the direct product of the transmittance through the covers, τ, and the collector plate absorptance, α, because some of the radiation originally reflected from the collector plate is reflected back again from the cover. The heat loss coefficient, U, rises very rapidly with increasing wind velocity if there are no covers, but is less dependent when the collector has at least one cover. The number and spacing of the covers and the conditions within the spaces can be significant, for example an evacuated space greatly reduces heat losses.

The three design factors, F, $(\tau\alpha)$ and U, define the thermal performance of the collector. The overall efficiency of the collector, $\eta = Q/G_c$, can then be expressed in terms of the temperature difference, $(T_m - T_a)$, and the incident solar radiation, G_c.

$$\eta = \frac{Q}{G_c} = F((\tau\alpha - \frac{U}{G_c}(T_m - T_a)) \tag{8.3}$$

The longwave radiation emitted from the collector plate surface can be considerably reduced by treating the collector surface. The treatment reduces its emissivity in the longwave spectrum without greatly reducing the absorptivity for shortwave radiation. Any collector surface with this treatment is known as a selective surface. Many methods have been used to achieve these surfaces, including chemically applied coatings.

Equation 8.3. can be used to compare the performance characteristics of different types of collector and to explore the effects of altering the various design parameters. This is illustrated in Figure 8.4. which describes the performance of six different types of collector.

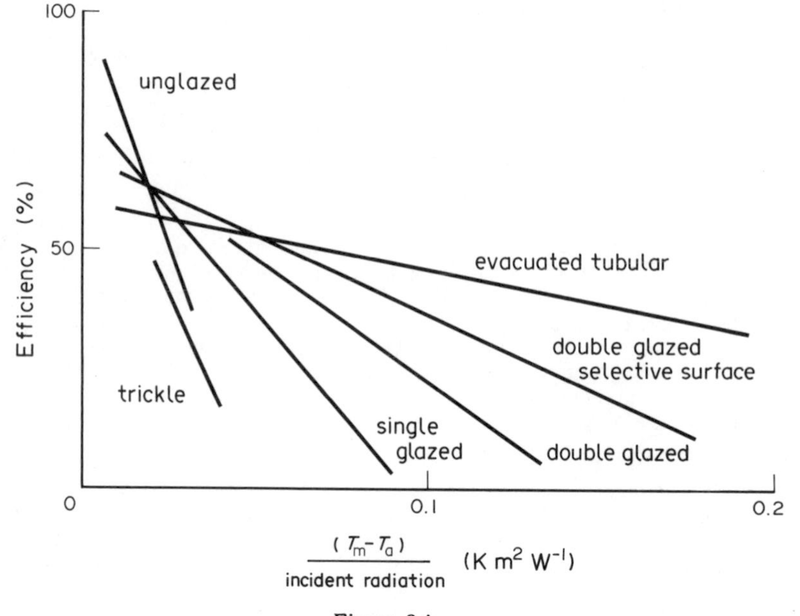

Figure 8.4

The unglazed collector performs well for small values of (T_m - T_a) and with good radiation conditions. A similar performance is obtained from the trickle collector, based on the use of black corrugated sheets with water flowing down the channels from a perforated pipe. For higher temperatures one or two transparent covers are needed and the use of a good selective surface also enhances the performance with lower incident radiation values. The best all round performance to provide temperatures above 60°C is given by the evacuated tubular collector, described later in the Thermal Power section.

A more detailed assessment with descriptions of many different types of collector has been carried out by the author[1]. This assessment includes details of international standards and testing procedures adopted by a number of countries.

By the early 1980s a solar water heating industry had been established in many of the western industrial countries. Its size depended on many factors, the main influences being the annual solar radiation level and the extent of government subsidies and support. Countries such as the United States, France, Germany, Italy and Japan had a substantial number of the larger multi-national companies involved in solar water heating, while solar industries were also being established in many of the Arab countries[1]. One of the most impressive urban solar water heating programmes in the world is in Romania, where the research team at the Building Research Institute in Bucharest has developed a number of systems, each with several thousand square metres of collector. Figure 8.5 shows a large collector system being completed in Bucharest. The collectors are seen on the roofs of the blocks of flats and the hot water storage tanks are in the foreground.

Figure 8.5

Space Heating

The possibility of at least partially heating a building by solar energy has been demonstrated on many occasions over the past forty years. The criteria originally suggested by Telkes[6] in 1949 for solving the problems of collection, storage and distribution of solar energy have been somewhat modified since then as experience has been gained with an increasing number of installations. In the early work, emphasis was placed on the collection of winter sunshine. This has been broadened to include the very substantial contribution that can be obtained from diffuse radiation. The three terms 'efficient', 'economic' and 'simple' used by Telkes in relation to the collector have always been the goal of sound engineering practice and an analysis of some of the earlier solar installations indicated that very few can satisfy all three.

The associated problem of storing the solar energy collected in the summer for use the following winter has attracted a tremendous research effort. The principle of using a very large, heavily insulated water storage tank buried underneath the building was tried over forty years ago at the first Massachusetts Institute of Technology solar house[7]. The comment that this approach was highly uneconomical had a considerable influence on the direction of storage system research over the next two decades. The effect of latitude and local radiation characteristics are now widely appreciated. It was originally thought that only a few days' storage could be economically viable so that the solar energy received during clear winter days would be made available for successive overcast periods and that this could only occur to any extent in parts of the world where there are appreciable amounts of winter sunshine. However, longer storage periods of up to several months were achieved in several solar houses during the mid-1970s and by the end of the decade interseasonal storage had been shown to be not only possible but also cost effective. For shorter term storage considerable reductions in the total volume of the store are possible by the use of chemical storage methods pioneered by Telkes.

The term 'solar house' first became familiar in the USA during the 1930s, when architects began to use large, south-facing windows to let the lower slanting rays of the winter sun penetrate into the back of the room[8]. At that time there was no real distinction made between the active systems, which use pumps or fans to circulate heat, and passive systems, which use solar energy naturally, involving the conventional building elements for solar energy collection, storage and distribution. The large windows saved fuel during the day, but it was not possible to store the solar energy. At night and during cloudy days the heat loss was so great that there was a relatively small saving on fuel during the entire heating season. This approach was later known as the 'direct gain' system.

Prior to 1973 there were probably less than 50 well-known solar houses or buildings in the world. Among these, Europe's largest, and for many years the world's largest passive building was the annexe to St. George's School, Wallasey[1], completed in 1962. By the end of 1977 Dr. W.A. Shurcliff, who had published a series of surveys on solar heated houses, confessed that the expansion in the number of solar houses had been so great that it was quite impossible to attempt to collect and cross-reference all the information[9]. There are now thousands of solar houses and buildings throughout the world. In parts of the United States and Northern Europe the early designs in the 1970s were all active systems, but the need for long term interseasonal storage was appreciated.

The Swedish group Studsvik Energiteknik AB developed a system based on cheap excavated and insulated pits, or heat storage tanks, with floating insulating covers. The solar collectors are mounted on the cover, which rotates to track the sun. Integrated units of this type could be connected to a normal district heating network, or to a low temperature heating system. The prototype plant commenced operating in February 1979. The volume of the heat store is 640 m^3 and the collector areas is 120 m^2. The system is designed to supply the annual heat consumption of a low temperature heated office building with a floor area of 500 m^2. At the end of its first season it supplied 97% of the total heat demand[10]. Another Swedish project uses deep ground heat storage by penetrating rock or clay to a depth of up to 100 m by vertical holes spaced up to 4 m apart[11]. Solar heat collected by very simple unglazed collectors is used to heat water which is circulated deep underground. These successful large interseasonal stores are generally not suitable for individual houses and as the first series of monitored test results from active space heating systems began to show performances well below predicated values the emphasis changed to passive applications.

In addition to the direct gain system, mentioned above, four other passive systems have been categorized[12] as follows:

(1) **The Trombe wall** or thermal storage wall in which the heat is stored in a wall which also absorbs the solar energy when it comes through the glazing.
(2) **The solar greenhouse**, in which the features of the solar thermal storage wall

and the direct gain approach can be combined by building a greenhouse on the south side of a building (in the northern hemisphere). This approach was attracting very considerable attention in the early 1980s.

(3) **The roof pond.** In this system a shallow pond or tank of water sits on a flat roof with its surface generally contained by a transparent plastic sheet. Movable insulation is essential because solar input during the summer months can be very large and it is needed to protect the building from heat losses in the winter. In the winter, the movable insulation covers the top of the pond at night and the heat from the pond can radiate down to the house. During the summer the insulation covers the top of the pond in the day and is removed at night when the heated water can radiate out to the colder night sky.

(4) **The natural convective loop.** In this system air is usually used as the heat transport medium and works in the same way as the classic thermosyphon water heating system[1].

A passive project believed to be the largest monitored group of direct gain houses in the world in 1984 is the Pennyland Demonstration Project at Milton Keynes. An estate of 177 houses has been laid out to minimize overshading. The estate is divided into two sections having different insulation levels. A further sub-division involves south-facing houses with the glazing split equally between the north and south faces, with the other houses having their main living areas and glazing concentrated on the south side. Both groups are being compared with houses at different orientations and also with a group of houses randomly orientated and overshaded on a nearby estate[13]. Preliminary results were very promising, with savings in the order of 40% being common.

Most solar houses need large collection areas. In some cases collectors can be placed on the south facing wall and windows can also be integrated within the glazing. An example illustrating this approach is shown in Figure 8.6, a small architect-designed two-bedroomed house in northern Europe. Originally designed as a natural circulation passive air-heating system, fans had to be installed later to assist in air circulation.

Figure 8.6

Thermal Power and other Thermal Applications

The earliest recorded example of a solar powered engine was reported by a French Professor, August Mouchot, in the first book to be published on solar energy, 'La Chaleur Solaire et Ses Applications Industrielles' in 1869[14]. Mouchot had apparently constructed a parabolic focusing mirror which he first used to drive a small steam engine in 1860. He subsequently received a patent for the French Government in 1861. By 1878 he had successfully demonstrated solar powered refrigeration, pumping and cooking. Several years earlier, in 1872, a solar water distillation plant had been built in Las Salinas, in Chile, with a total area of 4756 m^2 of glass[15]. This was the largest in the world for many years. Initially some 19 000 litres of fresh water were produced daily at a cost of approximately one quarter that of a conventional coal-fired boiler system. The whole system was self-sufficient in energy as the salty water was pumped from local wells by a windmill.

There has always been a major need for water pumping in arid regions where clear sky conditions often give long periods of direct solar radiation. The early experiments led to the most spectacular solar engine development of the time - the Shuman-Boys Sun-Heat Absorber at Meadi in Egypt in 1913. The absorber consisted of five large parabolic mirror sections each 62.5 m long and 4.1 m wide between the edges of the mirrors, giving a total collecting area of 1277 m^2. Although the maximum recorded pumping horsepower was only 19.1 hp, this was attributed to the poor performance of the steam engine and pumping plant[16]. Using the same steam conditions a modern low pressure steam turbine could produce about 100 hp. After these technically promising early results, very little more work was carried out on solar thermal applications for the next forty years. Power generation from coal and oil became widespread and relatively inexpensive.

Overall solar thermal system efficiency

System efficiency is proportional to the product of the collector efficiency, the ideal Carnot efficiency (equation 1.10 page 6) and a factor relating actual engine efficiency to the Carnot efficiency. There is clearly a conflict between the desired collector temperature for maximum collector efficiency and the desired temperature for high Carnot efficiencies for any given ambient temperature. High collector efficiencies require the maximum temperature to be as low as possible, while high Carnot efficiencies need as high a maximum cycle temperature as possible.

Example 8.1 The efficiency of a solar collector is given by the equation

$$\eta = 0.78 - \frac{7.7\,(T_1 - T_2)}{G_C}$$

where T_1, is the water temperature at collector outlet, T_2 is ambient temperature and G_C is the incident solar radiation. The hot water from the collector is used as the constant temperature heat source for a vapour turbine with an efficiency which is 40% of the Carnot efficiency between the temperatures T_1 and T_2. The incident radiation is 850 Wm^{-2} and the ambient temperature T_2 is 15°C. The turbine and the collector are to operate with T_1 at 60°C. Determine

(a) The theoretical Carnot efficiency
(b) The overall system efficiency
(c) The maximum possible collector efficiency if T_1 is allowed to vary, keeping T_2 and G_C constant.

Solution: (a) $T_1 = 273 + 60 = 333$
$T_2 = 273 + 15 = 288$

$$\text{Carnot efficiency} = \frac{T_1 - T_2}{T_1} \times 100\% = \frac{45}{333} \times 100 = 13.51\%$$

(b) Collector efficiency when $T_1 = 60°$

$$\text{is } 0.78 - \frac{7.7(333-288)}{850} = 0.3724 \text{ or } 37.24\%$$

$$\text{system efficiency} = 0.1351 \times 0.3724 \times 0.4 = 0.0201 \text{ or } 2.01\%$$

(c) The maximum possible collector efficiency occurs when $T_1 = T_2$ and is 0.78 or 78%

Example 8.2 What collector area would be needed if the system in Example 8.1. were required to have an output of 10 kW? Is this a realistic system?

Solution:

$$\text{Collector area} \times \text{system efficiency} \times \frac{\text{incident radiation}}{1000} = 10 \text{ kW}$$

$$\text{Collector area} = \frac{10\,000}{0.0201 \times 850} = 585.3 \text{ m}^2$$

As the required collector area is relatively very large, it would be more sensible to choose a collector type which could give a similar efficiency at a much greater outlet temperature T_1, so that the Carnot efficiency, and thus the system efficiency, could be improved.

These examples show that unless a large area of relatively inexpensive collectors can be used, collectors capable of working at temperatures up to at least 250°C are essential for thermal power applications. These can be classified into two main classes, tracking and non-tracking focusing collectors, with a number of sub-groups in each class.

The two most important non-tracking focusing collectors are the compound parabolic concentrator (CPC) and the evacuated tubular collector. The CPC was developed in 1974 by Winston[17]. In cross-section it appears to take the form of a truncated parabola with a vertical axis. Winston's innovation was to place the focus of the parabola forming the right-hand section at the base of the left hand section and vice-versa. With careful choice of dimensions all the incoming direct radiation was concentrated on the absorber near the base. Concentration factors up to ten can be achieved without diurnal tracking. Evacuated tubular collectors use vacuum technology to surround the absorbing collector surface, often a selectively coated pipe, with a double walled evacuated glass tube. The focusing element plays a relatively minor role with this type. These collectors are now commercially available, with typical values of between 0.84 and 0.86 for $(\tau\alpha)$ and 1.5 W m^{-2} K^{-1}[18] for U.

Tracking focusing systems can be classified into three groups. Point focusing collectors usually have a tracking parabolic reflecting dish, in which the direct solar beam is reflected to a central collecting point fixed to the dish. Line focusing collectors have an absorbing pipe placed at the focus. They are either parabolic in cross-section or focused through a linear Fresnel lens, which consists of a series of wedge-shaped sections in a single flat unit, with each segment designed to concentrate the incident radiation onto the absorbing pipe. This type is often known as a trough collector. Concentration ratios of up to 40:1 can be achieved. For high temperatures and large-scale power generation, the central 'Power Tower' system is used. A large field of steered mirrors, or heliostats, reflect solar radiation to a single central receiver mounted on a tall tower. Concentration ratios approaching 3000:1 have been achieved with this type. The best example in the 1970s was the 1 MW solar furnace at Odeillo, built by the French Centre National de la Recherche Scientifique[19]. The parabolic mirror, which is 39.6 m high and 53.3 m wide, contains 9500 individual mirrors with a total reflecting area of 1920 m^2. It faces a field of 63 heliostats, with a total mirror area of 2839 m^2. During the 1970s studies in the United States[20] considered the design of individual units of 100 MWe capacity. The next major step towards this concept was the commissioning of a 10 MWe plant at Barstow, California, in January 1982. In Europe there are a number of smaller-scale complementary programmes supported by the EEC and the IEA. Their largest, known as Eurelios, was completed in Sicily for the EEC in 1981, and is shown in Figure 8.7. It has two types of heliostat with a total area of 7800 m^2 and a designed output steam temperature of 450°C. Europe's largest solar power tower system is the French Themis, rated at 2.5 MW and situated a few kilometres from Odeillo at Targasone at an altitude of 1700 metres. The main system consists of a group of 201 heliostats with a total area of 10,740 m^2, facing a receiver on the top of a 101.5 metre tower. There is also an auxiliary system of eleven parabolic mirror concentrators, each with a surface area of 75 m^2, to maintain the salt solution used as the heat transfer fluid above its solidification point of about 150°C. Problems were being experienced with the corrosive salt solution at high temperatures and with gusty wind conditions shattering some mirrors. By July 1984 it was producing between 7000 and 8000 kWh daily. The aim of all these programmes was to examine as many different systems as possible before proceeding to fully commercial systems.

Solar Ponds

A solar pond is a body of saline water in which the concentration increases with depth. In a natural pond when solar radiation heats the water below the surface the action of convection currents causes the heated water to rise to the surface and the pond temperature normally follows the mean temperature of the surroundings. A solar pond contains concentrations of dissolved salts which gradually increase with depth, causing the density of the water to increase towards the base of the pond, which is often black. Solar radiation penetrates to the base, heating the water at this lower level, but any convection currents are suppressed by the density gradient. Heat losses from the surface are reduced, compared with a natural pond, and the temperature at the bottom of the pond rises. While there are daily fluctuations in both ambient air temperature and in the upper water layers, the temperature at the bottom of the pond, where heat would be extracted, remains fairly uniform[21]. A solar pond is both a massive heat collector and heat storage system and, compared with a conventional collector and heat store, is relatively inexpensive. Other advantages are:

(i) problems which could occur with dirt settling on the surface and reducing collection efficiency are eliminated,
(ii) extracting energy from the pond is very straightforward as the lowest hot layer can be pumped to the power station and returned to the pond for reheating,
(iii) solar ponds can operate continuously throughout the year provided the storage capacity is chosen to match the demand,
(iv) a solar pond power system can - like a hydroelectric plant - provide peaks of power on demand.

The system is illustrated in Figure 8.8.

Figure 8.7

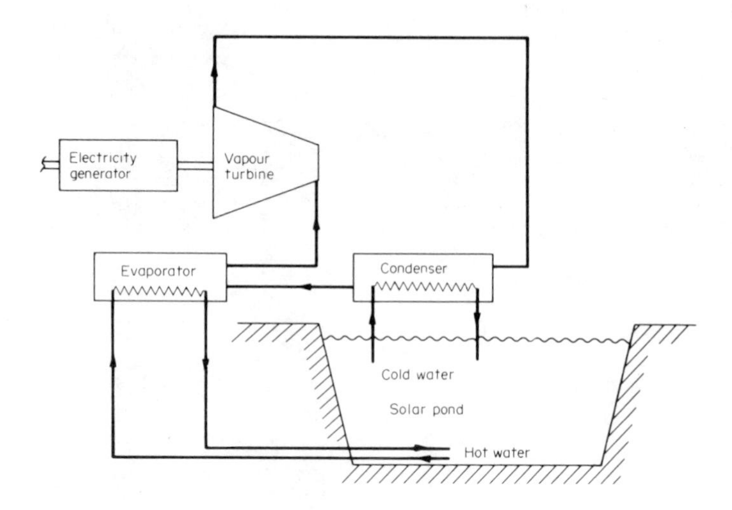

Figure 8.8

The analysis of its performance is complex and some early mathematical models were too simple. Wang and Akbarzadeh's analysis[22] is a good starting point for detailed study.

The development of the solar pond concept has been in a series of order-of-magnitude steps, with the few kilowatts of the early 1960s being followed by the 150 kWe solar pond at Ein Bokek in the Dead Sea area, developed by Dr. Harry Tabor of the Scientific Research Foundation in Israel. This was switched on at midnight on 16 December 1979 and has subsequently given outputs approaching 300 kWe. The third stage was the 5 MWe solar pond at Beth Ha'arava, also in the Dead Sea area. During the commissioning stage in 1983 temperatures above 86°C were achieved in March[23], and it is scheduled for a major demonstration in May 1984. Israel has plans to supply 2000 MW from solar ponds over a ten to twenty year period. The next major development will be a ten fold scaling up to a 48 MW power plant planned by the Southern California Edison Company[24]. This is based on the Israeli 5 MW plant and will be built in San Bernadino County by 1987. The Ormat Turbine Company will operate the plant, selling the power under contract for a thirty year period.

The range of possible applications continues to increase with space heating for buildings and greenhouses, process heating, space cooling, desalination and salt production already either in operation or under consideration[1].

Heliohydroelectric Power Generation

The concept of heliohydroelectric power generation is to convert solar energy into electricity by first transforming it into hydraulic energy. If a closed reservoir is completely sealed off from the sea, the level of the reservoir will tend to decrease as a result of evaporation. Hydroelectric generators could be placed at the reservoir end of pipes connecting the reservoir to the sea. The fall in the level of the reservoir induces a flow of water from the sea, and the potential energy caused by the difference in water levels could be transformed into electrical power. By choosing suitable water levels and power systems, it would be possible to have a continuous process. This topic has been extensively studied in Saudi Arabia by Kettani[25], who has measured evaporation rates and compared these with meteorological data. Kettani has also explored the possibility of building a dam across the Gulf of Bahrain, using the whole sealed-off Bay to create a hydraulic head from the open sea in the Gulf.

Connecting the Mediterranean Sea with the Qattara Depression for a hydroelectric scheme was first considered in 1916. This project was revived during the 1970s by a joint Egyptian-Federal Republic of Germany feasibility study. The results of this study, which was completed in December 1980, were reported in 1982 but a firm decision on whether or not to proceed with the scheme was not anticipated until the mid 1980s[26].

Distillation

One of the major problems in many parts of the world is the scarcity of fresh water and the development of inexpensive large solar distillation units capable of easy transportation and handling is increasingly important[1]. Solar distillation is another application which dates back to the nineteenth century and the simplest form of water still now in use is basically unchanged from the early designs which consisted of a shallow tray, filled with salt or brackish water, and covered by a sloping glass cover plate. The solar radiation heats the water in the tray and evaporates it. When the vapour comes in contact with the colder surface of the glass it condenses, forming fresh water which runs down the inner surface in the form of droplets and can be collected in a trough at the lower edge. Under good radiation conditions an output of about 4 litres of fresh water can be obtained daily per square metre of cover.

Australia has gained considerable practical experience over many years and a guide to the design, construction and installation of a solar still developed by the Commonwealth Scientific and Industrial Research Organization was published in 1965[27]. One of the best known large installations in recent years, with an evaporating surface area of 8667 m^2, was completed on the island of Patmos in the Aegean in 1967[28]. The average distillation rate was 3.0 kg m^{-2} per day, with a maximum of 6.2 kg m^{-2} at mid-summer. The first large installation designed and manufactured in the UK was a 185 m^2 unit for Aldabra in the Indian Ocean[29] in 1970. The development of various methods of improving simple still designs, such as the use of stepped salt water shelves or wicks, up to the early 1980s has been reviewed[1]. Later developments include the use of an 8 kW (peak) photovoltaic array to provide power for a reverse osmosis water desalination unit for a community of 250 people near Jeddah, Saudi Arabia[30].

Ocean Thermal Energy

The use of the temperature difference between the surface of the ocean and the colder deep water to operate a heat engine was proposed towards the end of the nineteenth century. The oceans are natural solar energy collectors and require no special storage systems or manufactured collectors and, because of their enormous size, have considerable potential to compete economically against other methods of power generation. The earliest system was a 22kW power plant off the coast of Cuba developed by Claude[31] in the late 1920s. It had an overall efficiency of less than 1% and operated with an open Rankine cycle in which the higher temperature sea water was passed directly to a low pressure evaporator to provide steam to power the turbine. It was uneconomic at that time, as was a subsequent larger project by the French some twenty years later, and further work was discontinued.

Renewed interest in the concept came in the 1960s when the possibility of using a closed Rankine cycle was suggested[32] in the USA. This work formed the basis for several major large-scale theoretical investigations which showed very similar overall net cycle efficiencies, ranging from 2.1 to 2.4%. Following these systems feasibility studies, a joint project called Mini-OTEC was developed by the State of Hawaii, Lockheed Missiles and Space Company and the Dillingham Corporation. Mini-OTEC was a 50kW (18kW net) plant which operated off the coast of Hawaii during 1979 to establish that the problems raised in the feasibility studies were surmountable. It was also intended to draw public and political attention to the OTEC system in contrast to the better-known solar technologies. The project was described by Henry J. White[33] who concluded that all the goals had been achieved and that a combination of private industry with enlightened state support could design, construct, test and operate a sophisticated energy system at a cost, and in a time scale, which a more formalized procedure might be unable to attain.

Among the advantages listed for OTEC by Avery and Dugger of the Ocean Energy Programme at the Johns Hopkins University[34] were

(i) it can deliver energy to regions bordered by warm ocean waters via direct electric-power transmission.
(ii) it can deliver energy to all other regions via an energy-intensive material such as ammonia, produced on OTEC plant ships in the tropics.

Ammonia was considered to be an outstanding choice as it could save the natural gas already used in conventional ammonia plants, it could serve as a synthetic fuel or it could provide an easily transported and storable source of hydrogen for fuel cells to generate electric power. The costs of OTEC electricity delivered directly or via fuel cells were projected to be competitive with costs from coal or nuclear plants by 1990 if the rapid OTEC development were to be pursued.

By the start of the 1980s, several other countries apart from the USA had active OTEC programmes. These included France, Japan and Eurocean, a European consortium[35]. A fascinating modification of OTEC technology was suggested by Professor Estefan of the National Research Centre, Cairo[36]. He pointed out that the Red Sea contains some zones of the hottest and saltiest water in the world within its volume of approximately 21 500 km^3. Below a transition zone between 100 and 400 m deep and above the deeper hot brines, the water temperature is reasonably uniform at 22°C. The intermediate brines and deep brines at levels of some 2000 m have temperatures ranging from about 44 to 56.5°C. Professor Estefan suggested that it appeared to be technically feasible to use an 'inverted' OTEC technique to utilize this temperature gradient. The electricity which could be generated from 30 000 OTEC plants of 325 MW (net) is about ten times greater than the estimated total world electric-power production in 1980[34].

The main obstacles to the rapid development of OTEC systems are those facing a number of other interesting large-scale renewable energy systems. There is clearly a concern that some of the environmental problems may not have been fully appreciated. The enormous cost of the development of full-size systems in the order of 400 MW could only be provided by government funding and there were no signs that this was likely in the early 1980s.

Refrigeration and Cooling

The great advantage of using solar energy in refrigeration and cooling applications is that the maximum amount of solar energy is available at the point of maximum demand. There are two quite different major applications, the first in the cooling of buildings and the second in refrigeration for food preservation or for storing vaccines for medical purposes.

In building applications, solar collectors can be used to provide a high temperature input to cooling systems using solar-powered engines or absorption systems. This part of the system could be used to provide heating outside the hot mid-summer period and the cost of the system could be shared between the two functions. The cooling demand is also at a maximum during the early afternoon, depending on the orientation of the building and its thermal mass, so that the storage capacity for cooling is only a few hours in contrast to the very much greater periods required for heating systems.

Solar powered refrigeration was first demonstrated over a hundred years ago[1] and several different systems based on the familiar absorption refrigerator were commercially available in the early 1980s. The four main methods used for refrigeration and cooling systems are as follows:

- (i) the vapour compression refrigeration cycle
- (ii) absorption systems
- (iii) evaporative cooling
- (iv) radiative cooling

In its simplest theoretical form the well-known domestic refrigerator can be thought of as a reversed heat engine, described in Chapter 1, page 6. In practice nearly all vapour compression refrigerators are driven by electric motors. Although small solar powered engines have been used to replace the electric motor in laboratory trials, it has proved to be too expensive to proceed to commercial applications. For future small-scale applications it is likely that conventional electrically powered refrigerators will be driven by photovoltaic cells through a d-c to a-c inverter, a method which is already proving to be economic in some applications.

There are two main types of absorption cooling systems. For continuous operation, the cycle is illustrated in Figure 8.9. The working fluid is a solution of refrigerant and absorbent, the most common being ammonia-water or water-lithium-bromide. When solar heating is supplied to the generator some refrigerant is vaporized and a weak mixture is left behind. The vapour is then condensed and expands to the lower pressure evaporator, where it is vaporized and refrigerates the external working fluid, which would be air for an air conditioning application. The cycle is completed in the absorber when the refrigerant recombines with the original solution and is pumped back to the generator. By the end of 1983 one of the largest systems in the world, the solar air conditioning system at the Chaim Sheba Medical Centre in Israel had been operating successfully for three years[37]. The system is powered by approximately 3000 m^2 of flat-plate solar collectors and annual savings of electricity are about 120 kWh per square metre of collector.

Intermittent ammonia-water absorption units are relatively simple to operate and have been successfully developed at the Asian Institute of Technology (AIT), Bangkok, by Professor R.H.B. Exell[38]. During the day, heat from solar collectors vapourizes the ammonia and it condenses in the condenser. At night the ammonia is passed through an expansion valve into an evaporator, where the refrigeration effect produces ice. By 1981 the principal unit at the AIT could produce up to 30 kg of ice from water at 28°C, with a collector area of 5 m^2 linked to the ammonia-water system. The solar radiation was enhanced by the use of plane reflecting mirrors. Two years later a much larger unit had been developed for use in rural village communities.

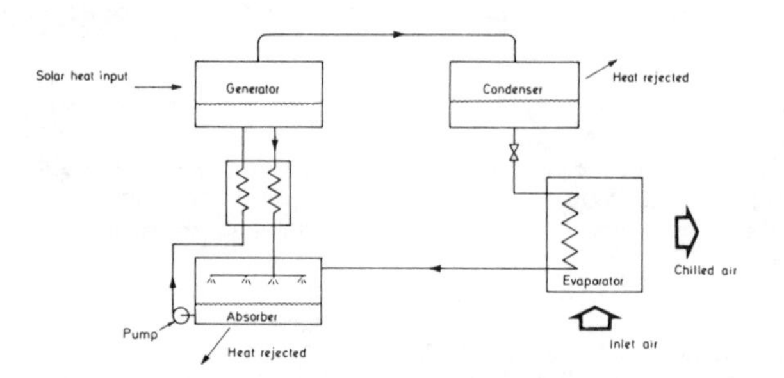

Figure 8.9

Evaporative systems achieve a cooling effect through the evaporation of water. Evaporative cooling can be achieved without making the air within a building more humid by combining a vertical water heating panel with a horizontal shaded roof pond[39]. Heat is absorbed from the room as shown in Figure 8.10. The heated water rises to the roof pond where it is cooled by evaporative cooling. An earlier simple method described by Thomason[40] takes water from a house storage tank and trickles it down an unglazed northfacing (in the northern hemisphere) roof.
Radiative cooling to the sky is possible at night with clear atmospheric conditions. Several examples of the use of the technique of pumping water into various roof-mounted collectors have been described[1].

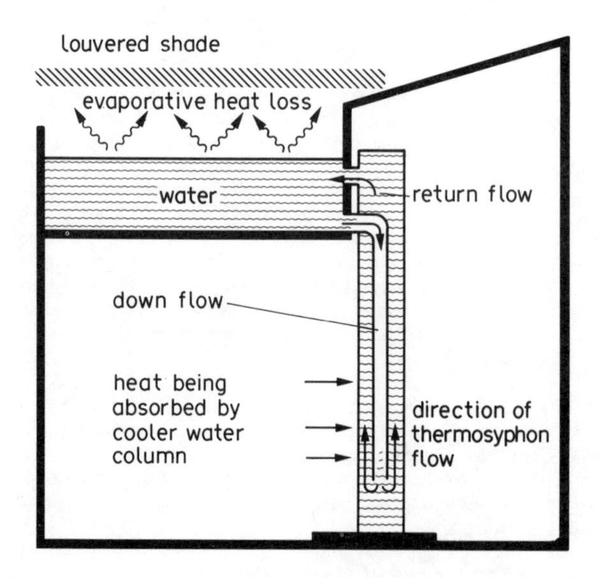

Figure 8.10

Photovoltaic Cells

The direct conversion of solar energy into electrical energy has been studied since the end of the nineteenth century. The early work was concerned with thermocouples of various different alloys, and efficiencies were very low, usually less than 1%. This work was reviewed by Telkes[41] in a paper written in 1953 and at that time it was felt that little more could be achieved and that efficiencies in this order were quite unsuitable for the generation of electricity. A similar pessimistic view was expressed in the UK[42]. However, in 1954 the Bell Telephone Laboratories in the USA discovered that thin slices of silicon, when doped with certain traces of impurities became a factor of ten times or more efficient at the conversion of solar radiation to electricity than the traditional light sensitive materials used in earlier photocells. Since then there has been a steady history of improvement and considerably higher conversion efficiencies have been quoted - up to 19% for silicon cells and over 20% for certain new gallium arsenide cells under laboratory conditions. The use of solar cells in space applications is well known and the development for terrestrial applications has accelerated so rapidly over the past few years that it is now believed that solar cells could provide a small but significant proportion of the electrical energy requirements of many countries throughout the world by the end of the century.

Among the advantages listed for the modern solar cell are that it has no moving parts to wear out, has an indefinitely long life, requires little or no maintenance and is non-polluting[43]. Unlike other types of electrical generator it is suitable for a wide range of power applications from less than a watt to several thousand megawatts. Although a Japanese estimate considered that a 10 MW generating station made with the technology available in 1974 would require the entire world production of silicon, about 1000 tonnes at that time, a more recent estimate in 1981[44] suggests that amorphous silicon, cast polycrystalline silicon and cadmium sulphide are free of serious problems of material supply.

Types of Solar Cell

The doping of a very pure semiconductor with small traces of impurities can modify its electrical properties, producing two basic types: p-type, having fixed negative and free positive charges, and n-type, having fixed positive and free negative charges. If these two types are placed together and the surface is exposed to sunlight, electrons will diffuse through the p-n junction in opposite directions giving rise to an electric current, as shown in Fig. 8.11. The earliest solar cells were made of silicon and one type of modern silicon cell is made by doping a slice cut from a single crystal of highly purified silicon with phosphorous, arsenic or antimony and diffusing boron into the upper surface, forming a 'p-on-n' cell.

An alternative method, forming a 'n-on-p' cell has proved to be more effective and is preferred. A detailed discussion of the physics of solar cells is beyond the scope of this text, but is widely available in the literature[45,46]. The front of the cell is protected by a transparent cover. This was usually made of thin glass or fused silica (quartz) for the early space applications, but a wide variety of other methods using weather resistant plastics had been adopted for many terrestrial applications[46].

The commercial production processes in the mid to late 1970s were complex. For example, during one stage, the 'pulling' of a crystal from the melt, temperature control within ±0.1°C at 1420°C was involved[47]. Consequently these early solar cells were expensive. When assembled in a group of encapsulated cells, known as modules, module costs were between US$25-90 per peak watt in 1975, expressed in 1975 dollars. A peak watt (Wp) is defined as one watt output at a light intensity of 1000 Wm^{-2} and a temperature of 25°C. The production of these very pure single silicon crystals, or electronic grade silicon, was a major factor in this high cost and it

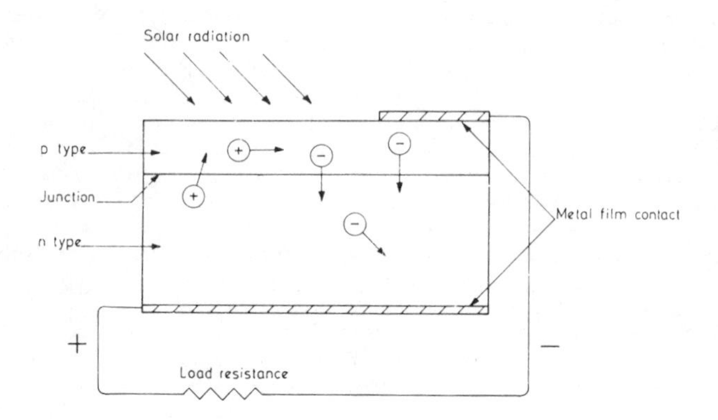

Figure 8.11

was hoped that a solar grade silicon could be developed. This would have more impurities than the electronic grade silicon, but would perform as well at less than one fifth of the cost. By the early 1980s however, this breakthrough had not been achieved and many other processes were under development. Amorphous, or uncrystallized, silicon can be deposited in very thin films or layers. This could lead to relatively inexpensive cells in the small power range, typically less than one peak watt, where their relatively low efficiency, in the order of 3-5%, would be of minor importance. Polycrystalline silicon, which consists of many small single crystals arranged at random has also been deposited in thin layers on a wide variety of materials by different methods such as fusion, dipping or coating. Practically all these processes give cells with laboratory efficiencies higher than 8%[48], with some greater than 15%. A radically different production technique has been investigated in several laboratories. By growing silicon crystals directly in ribbon form it is possible to eliminate the costly process of cutting very thin wafers of silicon from large cylinders of single crystal. One process, known as edge defined film-fed growth (EFG) was first investigated at the Tyco laboratories in the 1970s, but another process, the edge-stabilized ribbon (ESR) process developed by Arthur D. Little Inc. in 1981 as a result of research work carried out at Massachusetts Institute of Technology by Emanual Sachs, was considered to represent "a significant milestone in the development of photovoltaic materials to meet industry production goals"[49]. In the first results announced, ribbon lengths of up to 4 m and 25 mm wide, had been obtained at manufacturing rates from 25 to 150 mm per minute, with the possibility of continuously adjusting the thickness from 4 μm to well over 300 μm. Widths of over 100 mm were considered to be technically feasible. An important advantage over other ribbon growth methods which need the temperature to be controlled within one-tenth of a degree, is that the ESR process can perform reliably with temperature variations greater than 10°C.

A wide variety of other materials in addition to silicon are being considered for use in solar cells, with life expectancies of over twenty years predicated from accelerated life tests. Two types of cell with fairly similar properties are gallium arsenide and indium phosphide. Gallium arsenide has the ability to withstand considerable concentration and up to 1000 times full sunlight has been reported by the Plessey Company in the UK and Varian Associates in California. Research is also being sponsored into organic semiconductors and Schottky barriers (metal-to-semiconductor junctions).

By the end of 1983 although a number of promising research programmes had been cancelled or postponed, announcements of new developments or improvements in existing processes appeared regularly[50]. For example during December 1983 there were reports of a flexible amorphous silicon solar cell built on stainless steel 0.1 mm thick. A 10 cm^2 standard module of this cell, developed by Fuji Electronic, of Tokyo, was claimed to have attained 7.25% efficiency. Another Japanese company in Tokyo, Komatsu, had developed an amorphous silicon cell with 10.7% efficiency. This cell has a transparent electrode coated with indium tin oxide to enhance the light refraction angle and allow more light to reach the amorphous silicon layers. A third company, the Electrotechnical Laboratory at Tsukuba, near Tokyo, reported the development of a p-n junction film made from two kinds of pigment, both organic materials. In Australia, at the University of New South Wales, a new type of solar cell, the metal-insulator-semiconductor, with an efficiency of nearly 20% had also been developed without the normal p-n junction. In this case it was replaced by a fine metal grating placed on top of a thin layer of oxide which was allowed to grow on top of the silicon.

Applications

Although the present high cost of commercially available solar cells makes them unattractive in any situation where a conventional electricity supply is available, there are an increasing number of applications where solar cells are already economically competitive. One of the earliest applications was in the provision of unmanned lights at sea. Flashing lights on buoys, lighthouses and offshore oil rigs are increasingly being powered by solar cells and early applications were reported from the Gulf of Mexico, Japan and the UK[1]. The main problem in these earlier tests was the hostile marine environment and the salt in the atmosphere was found to atack some resins and plastic based mounting boards. Automatic weather stations and other remote instruments that are difficult to reach are now being considered for solar powered operation. Although the total costs using conventional fossil fuels may appear to be lower, they must also be offset against the cost of access for maintenance and refuelling.

By 1975 the world's largest terrestrial photovoltaic system was the solar array of 1 kW peak capacity installed by a commercial company in the USA, the Mitre Corporation[51]. Their energy storage system consisted of battery storage for short term and peak power requirements, and an electrolysis hydrogen gas system fuel cell combination for base load and operation at night.

The next six years saw a remarkable three-hundredfold increase in the maximum size of the solar array, with a 25 kW water-pumping system in Nebraska by 1977, a 60 kW system augmenting a grid system in California in 1979 and two systems of over 300 kW reported during 1981. One was a concentrating collector photovoltaic system to provide the Mississippi Community College in Arkansas[52] with both electricity and thermal energy. The peak electrical output was designed to be 320 kW from 59 400 single-crystal silicon cells, with parabolic trough concentrating collectors producing an average concentration of 30 suns by tracking the sun throughout the day. The 270 concentrator units occupy an area of about 1.5 ha, with a collector surface area of 3500 m^2 capable of producing about 7000 kWh thermal per day for winter heating. This is illustrated in Figure 8.12. A major problem with the installation has been the cost of maintaining the prototype solar cell system.

Figure 8.12

The other was a 350 kW system to supply the villages of Al-Aineh and Al-Jubaila in Saudia Arabia. The project formed part of the $100 million five-year joint programme between Saudi Arabia and the USA, known as SOLERAS. Other SOLARAS projects included several types of air-conditioning coolers and solar-powered desalination units.

During 1981 the Sacremento Municipal Utility District (SMUD) confirmed that the United States Department of Energy had been asked to underwrite a proposal to build a 100 MW photovoltaic power plant at the Rancho Seco nuclear power plant near Sacremento. It was planned to be built in stages over a period of twelve years with the first stage of 1 MW due for completion after 18 months' work. The final stage would be 35 MW, and when completed would show a further three-hundredfold increase in the maximum size of the solar array from the 1982 figures.

The use of solar cells for space applications is well known. This led to a proposal by Glaser in 1968[53] to use a satellite in orbit round the earth to produce electricity which could be fed to microwave generators for transmission to earth. The basic principle of the Satellite Solar Power Station is photovoltaic solar energy conversion, after concentration, from two symmetrically arranged solar cell arrays. The microwave generators form an antenna between the two arrays, which directs the mircowave beam to a receiving antenna on earth. In synchronous orbit the satellite would be stationary with respect to any particular location on earth and full use could be made of practically continuous solar radiation. Up to 15 times more energy is potentially available on this basis, compared with terrestrial applications which are limited by weather conditions and daily cycles. The system could be designed to generate from 3 to 15 GW[54].

A full assessment of the concept has been made in the USA by the Department of

Energy and NASA and, on a smaller scale, in Europe by the European Space Agency. By the early 1980s no single constraint had been identified which would preclude continuation of research and development for either technical, economic, environmental or societal reasons[55].

An estimate of world-wide installed photovoltaic capacity by the end of 1983 compiled by the Financial Times World Solar Markets team[50] suggested that the total capacity had reached about 10 000 kW peak. The major geographical locations were the United States (5000), Europe (1000), the Middle East (1000) and Japan (575). The main applications were classified as shown in Table 8.1.

Table 8.1

Worldwide installed photovoltaic capacity in 1983

		Kilowatts Peak
Public Sector Demonstration Projects (including village power and water)		3 750
Private Sector Demonstration Projects		1 400
Communications		1 100
Navigation and Signalling		400
Cathodic Protection		200
Water Pumping		400
Remote Housing		1 400
Consumer Products (calculators, watches, toys)		1 200
Other Applications		150
	Total	10 000

Data from reference 50.

The effects of inflation and fluctuating exchange rates make it difficult to assess any predictions of future costs, but by 1984 it was clear that targets set for 1985 some ten years earlier, which corresponded to about three US dollars (1984 dollars) per peak watt, would be within grasp.

The long-term future was very clearly outlined by Dr H.L. Durand, former President of the Commissariat à l'Energie Solaire in France in 1979[48]

> If photovoltaic modules remain above one US dollar (1979 dollars) per peak watt, a large industry can be built in order to replace the billion-dollar market of small and medium size power stations, a market which is held today by diesel generators. If the cost comes down below 0.5 US dollars/Wp the jump may become fantastic, since photovoltaics may then compete with the large fuel or nuclear stations, provided that the storage problem can be solved. The hopes certainly justify the large amount of effort and money which are currently allocated to this exciting challenge.

Some economic, legal and social issues

An analysis by the International Institute for Applied Systems Analysis, Austria, of the potential role of solar energy applications in Europe and the USA over the coming century showed that if new solar technologies were to provide 1% of US primary energy by 1985 and penetrated the US energy marketplace at a fractional rate substantially greater than that of oil or gas in the past, it would still require seventy years for solar energy to displace 50% of other primary sources[56]. This slow rate of penetration is consistent with the requirements for and availability of investment capital, including competition for this capital for other purposes such as housing, agriculture and new industrial facilities. Their work suggested that new institutional and economic mechanisms would be required to accelerate the embedding of solar technologies in national economies if these technologies were to make a significant contribution to energy needs by the early part of the next century.

One of the conclusions reached in a study[57] of the effect of solar space heating and domestic hot water systems on the new housing market during the period 1980-2000 was that the solar approach enabled 5000 new jobs to be created per million tonnes of domestic fuel oil saved, compared with the creation of only 600 new jobs by nuclear generated electric heating systems. The Institut de Sciences, Mathematiques et Economiques Appliquées in Paris drew a careful distinction between the supply or demand factors as main influences on economic growth. If investment is determined by supply, the decisive factor in choosing the technical alternative is competitiveness, often accompanied by maximum productivity. But if investment is determined by demand with the market protected from foreign competition, the technical choice can be oriented towards those technologies which have, at equal cost, the highest job requirements. The different multiplier effects show that if one job is created by the nuclear-powered electric heating systems, 2-2.5 jobs could be created by developing solar energy systems and heat pumps.

The problem which could face anyone who has installed a solar system of any type, active, passive, hybrid or photovoltaic, is simply stated but difficult to solve. What can be done if a neighbour builds a new dwelling or extends an existing building or allows a bank of trees to grow on his land, thereby shading your collectors? In an excellent analysis containing 17 references to this issue, Eisenstadt and Utton[58] point out that at least three pertinent questions arise:

(i) does the homeowner have a right to the sunshine that is blocked by his neighbour's building or tree?
(ii) if he does not have a right to the solar energy, should such a right be given him?
(iii) if such a right is given, how should it be done?

The answers to these questions will naturally vary from one country to another and may need careful assessment.

Summary

In the ten years since the events of 1973 there have been many major research, development and demonstration programmes in the applications of solar energy. A major constraint has been that solar systems must be seen to be competitive with conventional equipment before substantial market penetration can be achieved. Three main areas have been identified where this has already happened in a number of countries - solar water heating, passive systems for space heating and some photovoltaic applications. The future for photovoltaic applications appears to be so promising that both government and private sector funding for various solar thermal power applications has been reduced, with the exception of work on solar ponds.

Following the political decisions to invest in solar research, development and demonstration programmes, solar educational programmes at all levels could be the key to public acceptance that more government funding should be made available both for the proven solar applications and for the long term developments which may become the most viable, non-polluting and economic energy systems of the next century.

References

1. McVeigh, J.C., Sun Power, Second Edition, Pergamon Press, 1983.

2. Sayigh, A.A.M., Estimation of total radiation intensity, Jl Engng Sc, University of Riyadh, Saudi Arabia, 5,1,43-55, 1979.

3. Holmes, M.J. and Hitchin, E.R. An "example year" for the calculation of energy demand in buildings, Building Services Engineer, 43, 186-189, 1978.

4. Yellott, J.I., Solar energy utilization for heating and cooling, ASHRAE Guide and Data Book series, 59, 1974.

5. Smith, C.T. and Weiss, T.A., Design applications of the Hottel-Whillier-Bliss equation, Solar Energy, 19, 109-113, 1977.

6. Telkes, M., A review of solar house heating, Heating and Ventilating, 46, 68-74, September 1949.

7. Hottel, H.C. and Woertz, B.B., The performance of flat-plate solar-heat collectors, Trans ASME, 64, 91-104, 1942.

8. Nemethy, A., Heated by the sun, American Artisan, Residential Air Conditioning Section, August 1949.

9. Shurcliff, W.A., Solar Heated Buildings - a brief survey, 19 Appleton St., Cambridge, MA. 02138, USA., 1977.

10. Roseen, R.A. and Perevs, B.O., Integral seasonal storage solar heating plant, Solar World Forum, 1, 697-702, Pergamon Press, Oxford, 1982.

11. Platell, O.B., The sunstore-deep ground heat storage, low temperature collectors and indoor heaters, Solar World Forum, 1, 777-785, Pergamon Press, Oxford, 1982.

12. Stromberg, R.F. and Woodall, S.O., Passive Solar Buildings: A Compilation of Data and Results, SAND 77-1204, Sandia Laboratories, Alberquerque, New Mexico, 1977.

13. Chatfield, J., The Pennyland Demonstration Project, Solar World Forum, 3, 1823-1827, Pergamon Press, Oxford, 1982.

14. Mouchot, August, La Chaleur Solair et Ses Applications Industrielles, Gauthier-Villars, Paris, 1869.

15. Harding, J., Apparatus for Solar Distillation, Paper no. 1933, Selected Papers, Institution of Civil Engineers, 73, 1908.

16. Ackerman, A.S.E., The utilisation of solar energy, Trans. Soc. of Engs., 81-165, 1914.

17. Winston, R., Principles of solar collectors of a novel design, Solar Energy 16, 89-95, 1974.

18. Horster, H and Kersten, R., Evacuated Solar Collectors, UK ISES Conference (C25) on Recent Developments in Solar Collector Design, London, 1981.

19. Trombe, F., et. al, First results obtained with the 1000kW solar furnace, Solar Energy 15, 63-66, 1973.

20. Blake, F.A., 100 MWe solar power plant design configuration and performance, NSF-RANN Grant No. AER-74-07570, Martin Marietta Aerospace, Denver, 1975.

21. Saulnier, B., et. al., Field testing of a solar pond, ISES Congress, Los Angeles, Extended Abstracts, Paper 35/1, July 1975.

22. Wang, Y.F., and Akbarzadeh, A., A parametric study on solar ponds, Solar Energy, 30, 6, 555-562, 1983.

23. Jesch, L.F., Report on Solar World Congress, Perth, Sun at Work in Britain, 17, November, 1983.

24. News item, Electrical Review International, 1, 1, 1984.

25. Kettani, M.A. and Gonsalves, L.M., Heliohydroelectric power generation, Solar Energy, 14, 1972.

26. Abdelfattah, A., The Qattara hydrosolar project, Solar World Forum, 4, 3062-3069, Pergamon Press, Oxford, 1982.

27. Read, W.R.W., A solar still for water desalination, Report ED 9, CSIRO, Melbourne, 1965.

28. Aegean Island installs world's largest solar distillation plant, Civil Engineering and Public Works Review, 1005, September, 1967.

29. Porteous, A., Fresh water for Aldabra, Engineering, 490, 15 May 1970.

30. Crutcher, J.L., et. al., A stand-alone seawater desalting system powered by an 8 kW ribbon photovoltaic array, Solar World Forum, 2, 1110 - 1119, Pergamon Press, Oxford, 1982.

31. Claude, G., Power from the tropical sea, Mechanical Engineering, 52, 1039-1044, December 1930.

32. Anderson, J.H. and Anderson, J.H. Jnr., Large-scale sea thermal power, ASME Paper No. 65-WA/Sol-6, December, 1965.

33. White, H.J., Mini-OTEC, International Journal of Ambient Energy, 1, 2, 75-88, 1980.

34. Avery, W.H., and Dugger, G.L., Contribution of ocean thermal energy conversion to world energy needs, International Journal of Ambient Energy, 1, 3, 177-190, 1980.

35. Expanded Abstracts of the 7th Ocean Energy Conference, Washington DC, June 1980.

36. Estefan, S.F., Modified OTEC technology for the Red Sea area, Paper E6-8:91, ISES Solar World Forum Abstracts, Pergamon Press, Oxford, 1981.

37. Solar absorption chiller and total climate control system, Publicity Unit, Israel Export Institute, PO Box 50084, Tel Aviv 61500, Israel, 1984.

38. Excell, R.H.B., and Kornsakoo, S., The development of a solar-powered refrigerator for remote villages, Solar World Forum, 2, 1049-1053, Pergamon Press, Oxford, 1982.

39. Crowther, K., and Melzer, B., The Thermosyphoning cool pool: A natural cooling system, Proc. Third National Passive Solar Conference, San Jose, California, 3, 448-451, 1979.

40. Thomason, H.E. and Thomason, H.J.L., Solar houses/heating and cooling progress report, Solar Energy, 15, 27-40, 1973.

41. Telkes, M., Solar thermoelectric generators, J. Appl. Phys. 25, 765-777, 1954.

42. Utilization of Solar Energy, Report of the NPL Committee (UK), published in Research, 5, 522-529, 1952.

43. Solar Energy: a UK assessment, UK Section, ISES, London, 1976.

44. Treble, F.C., Rapporteur's Report, Int. Conf. on Future Energy Concepts, Institution of Electrical Engineers, London, 1981.

45. Wilson, J.I.B., Solar Energy, Wykeham, London, 1979.

46. Rauschenbach, H.S., Solar Cell Array Design Handbook, Van Nostrand Rheinhold, New York, 1980.

47. Currin, C.C., et al, Feasibility of low cost silicon solar cells, 9th Photovoltaic Specialists Conference, Maryland, May 1972.

48. Durand, H., Present status and prospects of photovoltaic solar energy conversion, UK ISES Conference (C21) on Photovoltaic Solar Energy Conversion, London, 1979.

49. Glaser, P.E., in Arthur D. Little Inc.News, 24 October, 1981.

50. World Solar Markets, Financial Times, Bracken House, 10 Cannon Street, London EC4P 4BY.

51. Haas, G.M. et. al., Experience to date with the Mitre Terrestial Photovoltaic Energy System, ISES Congress, Los Angeles, Extended Abstracts, Paper 21/3, July 1975.

52. A Total Photovoltaic Energy System, Mississippi Community College, Blytheville, Arkansas, 1981, and Dr Harry Smith, personal communication, 1983.

53. Glaser, P.E., Power from the sun: its future, Science, 162, 857-861, November 1968.

54. Glaser, P.E., The case for solar energy, Conf. Energy and Humanity, Queen Mary College, London, September 1972.

55. Kessing, D. A survey of technologies required in the development of solar power satellites, Solar World Forum, 4, 3086-3094, Pergamon Press, Oxford, 1982.

56. Nakicenovic, N. and Weingart, J.M., Market penetration dynamics and constraints on the large-scale use of solar energy technologies, Paper J1:3, ISES Solar World Forum Abstracts, Pergamon Press, Oxford, 1981.

57. Outrequin, P., Solar energy and economic growth, Solar World Forum, 2, 1395-1403, Pergamon Press, Oxford, 1982.

58. Eisenstadt, M.M. and Utton, A.E., On the right to sunshine, Solar Energy, 20, 87-88, 1978.

Further Reading

A considerably more detailed overview of all the major solar applications has been published by the author in Sun Power: An introduction to the application of solar energy, Second Edition, Pergamon Press, Oxford, 1983. The earlier historical applications are covered by Ken Butti and John Perlin in A Golden Thread, Cheshire Books and Van Nostrand Reinhold, 1980. Dr J.I.B. Wilson's Solar Energy, Wykeham Science Series, 1979, provides an introduction in depth to the photovoltaic applications, while professional engineers and designers will find H.S. Rauschenbach's Solar Cell Array Design Handbook, Van Nostrand Reinhold, New York, 1980, a practical design tool. Architects have praised Edward Mazria's Passive Solar Energy Book, Rodale Press, 1979 and for thermal applications the leading basic reference source is Solar Engineering of Thermal Processes by J.A. Duffie and W.A. Beckmann, John Wiley, New York, 1981. Pergamon Press publish a number of important journals in the field including Solar Energy, Sunworld and Energy. Authoratitive reviews of selected topics appear regularly in Solar Energy. For example, Dr Harry Tabor's record of his solar pond projects appears in Volume 27, 181-194, 1981. The proceedings of the International Solar Energy Society's world conferences are also published by Pergamon Press every two years and must form the basis of any serious solar library.

Exercises

1. What are the main differences in solar radiation availability between a northern European country, such as the UK, and any country in the Middle East? How has this affected government attitudes to research, development and demonstration programmes?

2. What are the main applications of solar energy today in (a) the UK and (b) the Arab countries. How might these change over the next twenty years?

3. The Shuman-Boys solar engine at Meadi pumped water with a maximum recorded output of 19.1 horse-power. If the solar radiation was 894 Wm^{-2} at that time, what was the overall energy conversion efficiency of the system? The total collection area was 1277 m^2.

4. The efficiency of a solar collector with constant incident radiation is given by the equation

$$\eta = a - b(T_1 - T_2)$$

where T_1 is the water temperature at collector outlet, T_2 is the ambient temperature and a and b are constants. The hot water from the collector is used as the constant temperature heat source for an ideal heat engine operating on the

Carnot cycle. Show that the maximum system efficiency occurs when

$$bT_1^2 = T_2(a + bT_2)$$

Determine the maximum system efficiency if a = 0.58, b = 0.001 and the ambient temperature, T_2, is 30°C.

5. A solar water heating system installed in London has a collector area of 4 m² and faces south inclined at 10° to the horizontal. Which of the following statements could be correct?
   (a) It could provide up to 80% of the hot water requirements of an average family of four.
   (b) Its annual output is unlikely to be greater than 7200 MJ.
   (c) It would be less expensive to install 4 m² of photovoltaic cells.
   (d) If 4 m² of photovoltaic cells were to be installed their maximum output would be less than one kilowatt.

6. Describe the operation of the earliest solar pump. How could these principles be modified for use today?

7. What are the basic features of the n-on-p solar cell? Describe the main small-scale (up to 1 kW) applications of solar cells and examine trends in total system cost per peak watt since 1980.

8. Distinguish between active and passive solar heating systems. What are the main advantages and disadvantages of the two systems?
   A passive solar heating system has been designed to save the equivalent of 2750 kWh of electricity annually at 5p per unit. It's cost was £1300 and the loan is to be paid off in twenty equal annual installments at a fixed interest rate of 8.5%. How does the annual repayment compare with the savings provided by the system?

9. Examine the present state-of-the-art of Ocean Thermal Energy Conversion systems. What are the main environmental concerns? How could these be overcome?

10. Compare the advantages and disadvantages of government funded subsidies proportional to the capital investment in a solar system with subsidies proportional to the energy saved by the solar system. Should everybody be asked to pay the full social cost of the energy he or she consumes?

Answers

3. 1.25%
4. 15.15%
5. (b) and (d)
8. The two costs are approximately equal at £137.50.

9. GEOTHERMAL ENERGY, WIND, TIDAL AND WAVE POWER

Introduction

The four energy sources discussed in this chapter made a negligible contribution to world energy production in the early 1980s, perhaps less than a quarter of one per cent between them. Geothermal energy is the best developed source and the one most capable of very considerable expansion over the next two decades, both for electricity generation and thermal applications. Windpower has been used for small-scale mechanical applications for hundreds of years and more recently for electricity generation where considerable research efforts are being made to develop machines capable of delivering several megawatts. Tidal power has a very limited potential, with only a few large sites capable of exploitation and one major tidal scheme in operation. Wave power is the least developed of the four, but the one which has attracted very considerable research interest in the UK because the theoretical potential resource is significant.

GEOTHERMAL ENERGY

Geothermal energy is thermal energy stored in the earth. Although the earth's heat can be regarded as an infinite source of energy, prolonged exploitation can exhaust a geothermal field. Geothermal energy is, therefore, not a renewable source of energy[1]. In the nineteenth century it was believed that the residual heat in the centre of the earth was the source of the natural geothermal phenomena such as hot springs with jets of steam that could be seen on the earth's surface[2]. It is now widely accepted that there are two heat sources. The first arises from radioactive decay and the geological evidence points strongly to potassium, uranium and thorium contained in the rocks that form the earth's crust[3]. The second comes from the mantle which lies below the crust and which may also contain small concentrations of radioactive elements. The crust is some 30 to 35 km thick below the continental land masses and the boundary between the crust and the mantle is known as the Mohorovicic seismic discontinuity, or Moho[2,3]. According to the plate tectonic theory the crust is not a solid shell but consists of rigid segments or plates which can move relative to each other over the mantle. Pressure builds up at plate boundaries and the resulting sudden movement results in earthquakes and promotes the movement of large masses of molten rock or magma upwards into the earth's crust, causing volcanic activity[3]. The thermal effects of the interaction between plates can extend several hundred kilometres from the boundaries. Major plate boundaries are well-known and indicate the areas where exploitation of earth heat would be most likely to be successful.

Measurements of temperatures taken in mines and boreholes penetrating into the crust show that, with the exception of a very shallow zone near the earth's surface, the temperature rises as the depth increases. The rate of increase, or thermal gradient, is between 20 and 30K per kilometre for non-volcanic regions, with a smaller increase for older rock[3] and a much higher increase near magma penetration. Most of the heat which reaches the earth's surface does so by conduction, but some heat is transfered by convection to the free water in the outer few kilometres of the crust. This can occur by the simple process of groundwater sinking through permeable rock such as the sandstones or limestones, or where the rock has been fractured, into the hotter regions and then circulating back towards the surface, or by the heating of the groundwater by igneous activity[3]. Temperatures of these hydrothermal fluids in the range from 100° to 200°C are common and in places have reached 400°C, conditions which can result in some of the water flashing (changing state from liquid to vapour with rapid reduction in pressure) into steam and appearing as hot springs or geysers[3]. The average value for terrestrial heat flow on the continental land masses is about 0.06 Wm^{-2}[4,5]. A slightly higher value, 0.063 Wm^{-2}, is also widely quoted in the literature[6,7], but in the exploited geothermal fields the heat flow carried to the surface by the fluids can be from 200 to 1700 times this value[7].

Geothermal Resources

Compared with the proved reserves and ultimate resources for the fossil fuels, which are published at regular intervals, estimates for both national and world geothermal heat resources must be regarded with considerable caution. Armstead[1] points out that it is of more immediate interest to have an approximate idea of the amount of geothermal energy that could be obtained under existing economic and operating conditions, but that any attempt to estimate this must be highly speculative. Many regions in the world have had no geothermal exploration and it is very difficult to place any confidence in much of the published work. However, as a starting point the World Energy Conference[8] produced as assessment which is given in Table 9.1. The table is presented in units of 10^{20}J.

Table 9.1

Estimates of Geothermal Resources for Electricity Generation (10^{20}J)

Resource base, taking into consideration the continental land masses to a depth of 3 km and a datum of 15°C	410,000
Of this resource base only 2% is assumed to be of adequate temperature for electricity generation	8,200
Assume that the overall recovery and conversion efficiency is about 2.2%	180
One fifth is convertible by existing technology	36

Data derived from references 1 and 8.

In the early 1980s total world primary energy consumption was approximately 3×10^{20}J and the primary energy equivalent of electricity generated was about a quarter of this figure. Table 9.1 suggests that the geothermal resource is some 150 times greater than the world annual production of electricity. Armstead commented that while the estimate may well be of the right order and is probably as much as can reasonably be expected because of the lack of data, it is 'suggestive of rather insecurely based guesswork'[1]. The World Energy Conference also gave a figure of 2.9×10^{24}J for the estimated recoverable thermal energy which was theoretically available for direct applications at lower temperatures. Again, this figure must be fairly speculative as it amounts to just over 7% of the total estimated resource base. When estimates of geothermal resources are made for individual countries the same reservations must be applied. The wide range of estimates which can be given emphasize the need for caution. For example a series of estimates for Japan ranged from 40 000 MWe for the next thousand years, representing some 35% of the total world potential, to 8 650 MWe[1].

Geothermal areas, fields and aquifers

The surface of the earth can be classified into three main areas as follows[1]:

(i) Non-thermal areas with temperature gradients between 10 and 40 K per kilometre of depth.
(ii) Semithermal areas with temperature gradients approaching 80 K per kilometre of depth.
(iii) Hyperthermal areas with considerably larger temperature gradients.

An important distinction must be made between a geothermal area and a geothermal field. Many thermal areas are associated with rock of low or zero permeability and cannot be exploited under existing economic and operating conditions. Geothermal fields contain the hot water or steam in permeable rock formations and a number of these are operating commercially. They can also be classified into three main types[1]:

(i) **Semithermal fields** which can produce hot water at temperatures up to 100°C from depths up to 2 km.
(ii) **Wet fields** which produce water under pressure at temperatures greater than 100°C. When this water reaches the surface, its pressure falls and some flashes into steam, the remainder being boiling water at atmospheric pressure.
(iii) **Dry fields** which produce dry saturated or superheated steam at pressures above atmospheric pressure.

Another source of useful hot water is the low grade aquifer, which can produce water up to a temperature of about 75°C by drilling to depths of between 1.5 and 2 km, corresponding to a temperature gradient of 30 K per kilometre. Low grade aquifers can be found in non-thermal areas but are only worth exploiting if they are located fairly close to an appropriate application, such as space heating in a town or city.

Thermal Applications

The earliest application of geothermal energy was the use of natural hot springs for bathing or medical purposes. One hot spring near Xian, which was the capital city of ancient China, has been used for over a thousand years and still attracts many bathers. The history of the use of geothermal energy for industrial applications probably started in the Larderello area of Tuscany, in Central Italy, in 1827. Thermal energy from the hot wells was used in the crystallization of boric acid, which was also obtained from the natural pools formed from condensed steam and rainwater[7] and a flourishing chemical industry developed there over the next hundred years. An increasing number of applications for space heating have been reported since the 1960s. Examples of two different types are the semi-thermal fields in Iceland and the low grade aquifer found in the Paris Basin. The main features of a typical geothermal district heating system are shown in Figure 9.1., based on information from reference 9.

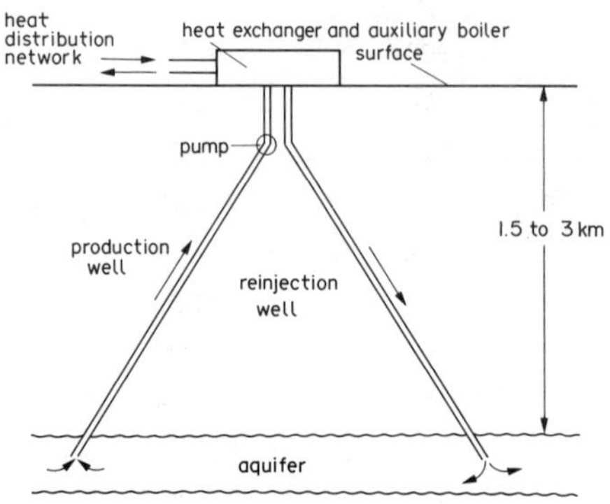

Figure 9.1

The first borehole establishes the characteristics of the aquifer and then becomes the production well, out of which the hot water is pumped. The second borehole is the reinjection well, which is used to dispose of the saline water after the heat has been extracted in the heat-exchanger. This well is approximately one kilometre from the production well to delay the return of cold water for 25 to 30 years. The auxiliary boiler can provide additional heat at periods of high demand and the whole system is connected to the housing units and buildings by a pipeline system. This must be no further than one kilometre from the geothermal wells, both for reasons of cost and prevention of heat loss. In Iceland the Reykjavik Municipal District Heating Services were able to sell hot water at less than one-fifth of the cost of heating with oil[10]. Running costs naturally depend on the methods of repaying the relatively high capital costs of each system. The geothermal plant installed in the Paris Basin cost approximately two thousand US dollars (1980 prices) per housing unit, which is comparable with a conventional system[10]. Among other thermal applications widely quoted in the literature are greenhouse heating including soil warming, drying of organic products, salt extraction and industrial process heating.

Electricity Generation

The history of electricity generation also started in the Larderello area, when a simple steam engine coupled to a DC generator was driven by steam form the geothermal field. This provided some electric lighting for the town of Larderello[7]. The steam engine was replaced by a 250 kW turbo-alternator in 1913. Until 1958 Italy was the only country where natural steam was used for power generation on an industrial scale. Production commenced in New Zealand in that year, followed by the Geysers field in the United States in 1960, when the total installed capacity in the world was 369 MW[4]. Development over the next decade to 697 MW by 1970 represented an annual growth rate of 6.5%, but the 1970s saw a considerable increase in growth. Table 9.2 shows the installed and projected geothermal electrical generating capacity for 1980, 1990 and 2000, taken from data presented at the United Nations Conference on New and Renewable Sources of Energy in 1981[10].

Table 9.2

Installed and projected geothermal electrical generating capacity(MW)

Country	1980	1990	2000
USA	923	4374	5824
Philippines	446	1225	1225+
Italy	440	560	800
New Zealand	202	282	382+
Japan	168	3668	3668+
Mexico	150	1000	4000
El Salvador	95	260	535
Iceland	32	68	68+
USSR	5	310	310+
Turkey	0.5	100	150
Indonesia	0.25	92	92+
Rest of World	-	88+	590+
Total	2462	12122+	17644+

+ Indicates that the figure is a minimum

Data from reference 10

Later data published by Shaw and Robinson[4] put the installed capacity in 1980 as 2082 MW. This gives a growth rate of 11.6% per annum during the 1970s. By examining the known orders for new plant they concluded that a realistic assessment of installed capacity in 1990 would be between 3786 MW and 5645 MW, less than half the figure suggested by the United Nations Conference[10]. Their projected figures of 3786 MW and 5645 MW for 1990 would represent average annual growth rates over the decade of 6.1% and 10.5% respectively, which seems reasonable compared with nearly 20% per annum to achieve the figure suggested by the United Nations Conference. When the 6.1% and 10.5% growth rates are extended to the end of the century the projected installed capacity would be in the range from 6804MW to 15336MW. This upper limit is fairly close to the lower limit given in Table 9.2.

The risks and problems associated with geothermal projects are not unlike those in searching for oil. The success rate of geothermal drilling, when measured by the proportion of wells which strike exploitable hot water or steam, is probably greater than that of oil drilling. However, the rewards are mush smaller and there are risks. For example in the United Kingdom the first major geothermal demonstration project, a scheme intended to heat a proposed office and shopping complex in the centre of Southampton, was abandoned early in 1984. The well was successfully drilled to a depth of 1675 m where the water temperature was 74°C, but pumping tests on the well produced a flow rate of only half the expected volume. An analysis of the tests indicated that the geothermal reservoir was considerably smaller than originally estimated and could only have a 15 to 20 year lifetime compared with an anticipated minimum of at least 25 years[11]. Another example occured some years earlier in Krafla, Iceland, when a volcanic intrusion partically destroyed the steam field and considerably reduced the output[4].

Hot dry rocks

Geophysicists have suggested that rock at temperatures of 200°C can be found at drillable depths, less than 10 km, over large regions of the earth's surface[4]. This has resulted in a number of major research projects in which deep holes are drilled into these hot rocks and a system of cracks is propagated between them[1,4,9,10]. Most of the research is aimed at establishing an optimum method for generating these fracture patterns. The basic technique uses hydraulic fracturing and the first successful tests were carried out at Los Alamos in the United States during the early 1970s[9]. A fracture system some 600 metres in daimeter was created between wells 3km deep and up to 4.5 MW was removed as heat during the initial test period of 2000 hours. This showed that the concept was valid. In the UK the Camborne School of Mines have extended the work in the United States by a more sophisticated approach to fracturing. They initiate the fracture system by explosives and then follow up with hydraulic fracturing. These experiments at a depth of 2 km have proved to be a major step towards commercial viability and the final stage will be the construction of a full-depth prototype extending to depths of 6 km[9]. The potentially exploitable granites in the south-west of England alone contain the equivalent of 8000 million tonnes of coal[12].

Some factors influencing developments

The economics of the applications of geothermal energy depend on the costs of competitive fuels. Where there are active geothermal fields and scarce indigenous resources, such as in Iceland, geothermal power is already the economic choice. Financial constraints and the lack of a suitable technical infrastructure can inhibit development in some of the poorer developing countries who would appear to have considerable hyperthermal field potential.
Several other possibilities for using geothermal energy have been discussed including the direct exploitation of the heat from active volcanoes[1,3,10]. This would have very considerable practical difficulties as it would involve tapping the magma at a depth of several thousand metres below the volcano, a technology which has not yet been developed. Among the ideas put forward for exploiting this source, Armstead[1] has suggested injecting water into the hot basaltic magma to produce hydrogen by dissociation.

The possible environmental problems which can arise from geothermal exploitation have been identified[1,9] and can include

- (i) The use of land for initial drilling operations and possible noise and damage.
- (ii) The long term visual impact and use of land for the power or heat extraction plant.
- (iii) The development of a suitable heat distribution and pipeline system.
- (iv) The release of gases, fluids and various chemicals during operation.
- (v) The physical effects on the geological structure of the area.

The earliest geothermal operations were carried out at a time when environmental issues were not taken into consideration. These early steam plants were reported to have unsightly tangles of steam transmission pipes, clouds of waste steam accompanied by a strong smell of hydrogen sulphide and, eventually, significant surface subsidence[9]. However in recent years these adverse effects, have been minimised. For example air pollution standards at the world's largest field, the Geysers in the United States with an installed capacity of 1300MWe in 1984, have resulted in 'cleaner' air than before the field was exploited. For the low aquifer systems the environmental impact should be negligible[9]. This has been confirmed by over ten years operating experience in France. The main problem is the disposal of the warm chemically-laden water after it has passed through the heat exchangers. As there are relatively few applications where this could be safely discharged into a well-mixed tidal estuary, it has become normal practice to reinject the brine back into the other end of the aquifer, as shown in Figure 9.1.

WIND POWER

Energy from the wind is derived from solar energy, as a small proportion of the total solar radiation reaching the earth causes movement in the atmosphere which appears as wind on the earth's surface[13]. The wind has been used as a source of power for thousands of years and the traditional horizontal axis tower mill for grinding corn, with sails supported by a large tower rather than a single post, had been developed by the beginning of the fourteenth century in several parts of Europe. Its use continued to expand until the middle of the nineteenth century when the spread of the steam engine as an alternative, cheaper, source of power started its decline. Nevertheless before the end of the nineteenth century several countries used the windmill as one of their main sources of power. In the Netherlands[14] there were about 10 000 windmills giving power outputs of up to 50 kW. In Denmark housemills were often mounted on the roofs of barns and, together with industrial mills, were estimated to be producing about 200 MW from over 30 000 units[15]. In the United States[14] an estimated six million small multi-bladed windmills for water pumping were manufactured between 1850 and 1940.

Work on the development of wind-generated electricity started in Denmark in 1890 when Professor P. La Cour obtained substantial support from the Danish Government, which not only enabled him to erect a windmill at Ashov, but provided a fully instrumented wind tunnel and laboratory. Between 1890 and his death in 1908, Professor La Cour developed a more efficient, faster running windwheel, incorporating a simplified means of speed control, and pioneered the generation of electricity. The Ashov windmill had four blades 22.85 m in diameter, mounted on a steel tower 24.38 m high. Power was transmitted, through a bevel gearing, to a vertical shaft which extended to a further set of bevels at ground level, and the drive was connected to two 9 kW generators - the first recorded instance of wind-generated electricity. By 1910 several hundred windmills of up to 25 kW capacity were supplying villages with electricity. The use of wind-generated electricity continued to increase in Denmark and during the 1939-1944 war period, a peak of 481 785 kWh was ontained from 88 windmills in January 1944[16]. The first windmill to feed electricity directly into a grid network was built by the Russians at Yalta, near the Black Sea[17]. Used as a supplementary power source, it was connected to a conventional fossil-fuel plant at Sevastopol, about 30 km away. It had three blades 30.48 m in diameter driving a 100 kW induction generator through wooden gears. The tower was 30.48 m high but was provided with an inclined strut to carry the thrust of the wind from the top of the tower to the ground. The base of the strut was driven round a circular track by an electric motor controlled by a wind direction sensing vane at the top of the tower. An annual output of 279 000 kWh was reported from the site which had an annual mean wind velocity of 6.7 ms^{-1}, but satisfactory control was difficult to achieve.

Large-scale modern windpower dates from the designs of an American engineer, Palmer C. Putnam[18] in the 1930s. He was responsible for the Smith-Putnam windmill which was erected at Grandpa's Knob in central Vermont in 1941. It had two blades with a diameter of 53.34 m and at that time it was the world's largest ever windmill, a record it was to hold for the next 35 years. The synchronous electric generator and rotor blades were mounted on a 33.54 m tower and electricity was fed directly into the Central Vermont Public Service Corporation network. The windmill was rated at

1.25 MW and worked well for about 18 months until a main bearing failed in the generator, a failure unconnected with the basic windmill design. It proved impossible to replace the bearing for over two years because of the war and during this period the blades were fixed in position and exposed to the full force of the wind. During the original assembly of the mainly stainless steel blades and supporting spars, rivet holes had been drilled and punched in the blades and cracks had been noticed in the metal around the punched holes in 1942. It was decided to carry out repairs on site, rather than returning the whole assembly to the factory. On 26 March 1945, less than a month after the bearing had been replaced, the cracks widened suddenly and a spar failed causing one of the blades to fly off. The S. Morgan Smith Company, who had undertaken the project, decided that they could not justify any further expenditure on it, apart from a feasibility study on the installation of other units in Vermont. This indicated that the capital cost per installed kilowatt would be some 60% greater than conventional systems.

Although sceptics have tended to regard this experiment as an expensive failure, it was the most significant advance in the history of windpower. For the first time synchronous generation of electricity had taken place and been delivered to a transmission grid. Both mechanical failures were due to a lack of knowledge of the mechanical properties of the materials at that time. Bearing design and the problems of fatigue in metals have been studied extensively since then and similar failures are unlikely to occur in modern windmills. Their research programme included an extensive series of on-site measurements, which proved that the actual site at Grandpa's Knob had a mean wind velocity of only 70% of the original estimated velocity and that many other sites should have been selected. The technical problems of converting wind energy into electricity had been largely overcome and the possibility of developing wind power as a national energy resource in any country with an appropriate wind climate has been established. However, very few wind turbines were to be built over the next thirty years.

Wind energy potential

Wind has a dependable annual statistical energy distribution but a complete analysis of how much energy is available from the wind in any particular location is rather complicated. It depends, for example, on the shape of the local landscape, the height of the windmill above ground level and the climatic cycle. Somewhat surprisingly, the British Isles have been studied more extensively than practically any other country in the world[19,20] and the west coast of Ireland, together with some of the western islands of Scotland, have the best wind conditions with mean average wind speeds approaching 9 ms^{-1}. The kinetic energy of a moving air stream per unit mass is $\frac{1}{2}V^2$ and the mass flow rate through a given cross-sectional area A is ρAV, where ρ is the density. The theoretical power available in the air stream is the product of these two terms

$$\tfrac{1}{2}\rho AV^3 \tag{9.1}$$

If the area A is circular, typically traced by rotor blades of diameter D, then

$\frac{\pi}{4}D^2 = A$, and the power available becomes

$$\tfrac{\pi}{8}\rho D^2V^3 \tag{9.2}$$

The actual power available can be conveniently expressed as

$$C\tfrac{1}{2}\rho AV^3 \tag{9.3}$$

where C is the coefficient of performance or power coefficient. The maximum amount of energy which could be extracted from a moving airstream was first shown by the German engineer Betz, in 1927, to be 16/27 or 0.59259 of the theoretical available power. This efficiency can only be approached by careful blade design, with blade-tip speeds a factor of six times the wind velocity and is known as the Betz limit. Modern designs of windmills for electricity generation operate with power coefficient values (C) of about 0.4, with the major losses caused by drag on the blades and the swirl imported to the air flow by the rotor[21]. Any aerogenerator will only operate between a certain minimum wind velocity, the starting velocity V_S, and its rated velocity V_R. Typically V_R/V_S lies between 2 and 3. If the pitch of the blades can be altered at velocities greater than V_R, the system should continue to operate at its rated output, the upper limit depending only on the design. In some systems the whole rotor is turned out of the wind to avoid damage at high wind speeds. An annual velocity duration curve for a continuously generating windmill is shown in Figure 9.2.

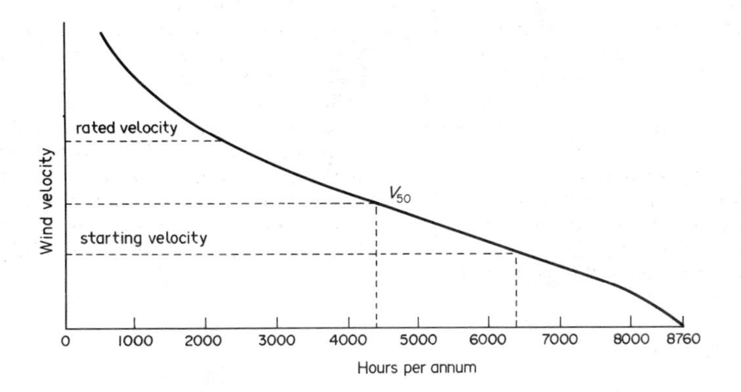

Figure 9.2

The effect of the height of the windmill tower on the performance can be significant and empirical power law indices have been established[22] relating the mean wind velocity V top the height H, in the equation $V = H^a$. A value of a = 0.17 is the accepted value in the UK for open, level ground, but this rises to 0.25 for an urban site and 0.33 for a city site. An ideal site is a long, gently sloping hill.

The mean annual wind velocity is normally used to describe the wind regime at any particular location, but the output from a windmill is proportional to V^3. Since a transient arithmetic increase in wind velocity will contribute much more energy to the rotor than an equal arithmetic decrease will deduct, the mean of V^3, which is always much greater than the cube of the mean annual wind velocity, should be used. For example, if the mean wind velocity is 8 ms^{-1} the most common variation in wind velocity occurs at frequent short intervals between 6 ms^{-1} and 10 ms^{-1}[15] and $8^3 = 512$, whereas $\frac{1}{2}(6^3 + 10^3) = 608$. A useful concept is the velocity exceeded for 50% of the year (4380 hours), shown in Figure 9.2. as V_{50}. This is quite close to the mean annual wind speed and has been used to give the annual extractable energy E_a if the rotor shaft is attached to an electrical generator as[23]

$$E_a = 3.2289 \, D^2 \, V_{50}{}^3 \text{ kWh} \qquad (9.4)$$

The Development of large wind-turbines

Details of the largest horizontal axis wind turbines built or planned in Europe since the late 1970s are given in Table 9.3. In 1979 the Twind windmill, located near the west coast of Scotland became the largest in the world since the Smith-Putnam machine[24].

Table 9.3.

Largest European Horizontal Axis Wind-Turbines

Country and Contractor	Rated Power (MW)	Hub Height(m)	Blade Diameter(m)	Number of Blades+
Twind, Denmark Private in 1979	2.0	54	54	3 D
Nibe A and B, Denmark Elsam, 1979	0.63	45	40	3 U
Growian I, FR Germany MAN, 1983	3.0	100	100	2 D
Growian II, Germany MBB	5.0	120	145	1 D
WTS2, Nasudden, Sweden KMW-ERND, 1982	2.0	80	75	2 U
WTS3, Maglarp, Sweden Swedyard/Hamilton Standard, 1982	3.0	80	78	2 D
Orkney, UK WEG	3.0	46	60	2 U

+ D represents a downwind orientation, U upwind

Data from references 13 and 21. The Growian II and Orkney machines were under construction during 1984.

A distinctive feature of the project was that it was not sponsored by the Danish government but resulted from a team effort from a college community who provided their own labour and finance. Normally its output is less than 1 MW and this is used to heat a large insulated 3000 m^3 hot water storage tank. The official Danish programme for large electricity producing wind energy systems started in 1977 with a joint programme directed by the Energy Ministry and the Electricity Utilities. Their major project has been the design, construction and operation of two machines, Nibe A and B, which were erected in 1979. These turbines are sited close to each other and are identical, apart from their rotor blades. Those for the A machine are supported by stays while the blades of the B machine are self-supporting.

In the UK two 100 kW machines were built in the early 1950s, the John Brown machine which was erected in the Orkneys and the Enfield-Andreau machine which was eventually erected in Algeria in 1957[13]. A wind data base was also established in the 1950s by the Electrical Research Association. Preliminary work with a design feasibility and cost study of large wind turbine generators suitable for network connection was carried out in 1976 and 1977 by a group comprising of British Aerospace Dynamics Group, Cleveland Bridge and Engineering Co Ltd, Electrical Research Association Ltd, North of Scotland Hydro-Electric Board, South of Scotland Electricity Board and Taylor Woodrow Construction Ltd[25]. A reference design was evolved for a 60 m diameter turbine in 1977. This eventually became the WEG (Wind Energy Group) 3MW design shown in Table 9.3., which is scheduled for completion in Spring 1985. Compared with all the other machines in Table 9.3., the WEG machine is rated for a much higher wind speed, 17.0 ms^{-1}. The other machines are rated between 12.0 and 13.0 ms^{-1}[27]. A smaller prototype, a 20 m diameter 250 kW machine, was commissioned in the summer of 1983.

The German wind programme, known as the Growian programme, had some 25 projects in operation during the early 1980s, ranging from some small, low-cost units rated at 15 kW for production in developing countries and a medium sized 25 m diameter twin-bladed 265 kW machine, the Voith-Hutter commissioned in 1981, to the large Growian I and Growian II machines.

The main feature of the Swedish programme was related to the design, construction and operation of two large-scale prototypes, located at Maglarp in the province of Skane in southern Sweden, and Nasudden on the island of Gotland. These projects aim at obtaining a basis for a decision in 1985 about the future of wind power in Sweden. A smaller experimental unit at Kalkugnen WTS 1, rated at 60 kW, gave valuable data before an accident occured during a planned change of rotor blades[13]. The successful collaboration between the Swedish group's WTS 3 at Maglarp and the American company Hamilton Standard resulted in an upgraded version, WTS 4, commencing operation at Medicine Bow, Wyoming, during 1983. WTS 4 is rated at 4 MW and is very similar in all its physical characteristics to WTS 3, but has a slightly higher rated wind speed, 13.7 ms^{-1} compared to 12.5 ms^{-1}.

In the United States the first major project in the official wind energy programme was the ERDA Model Zero (MOD-0) 100 kW windmill which consisted of a two-bladed, 38.10 m diameter, variable-pitch propeller system driving a synchronous alternator through a gearbox, mounted on a 30.48 m high steel tower[26]. The blades were located downstream from the tower and a powered gear control system replaced the traditional tail fin of earlier designs. This initial test programme was designed to establish a data base concerning the fabrication, performance, operating and economic characteristics of propeller-type wind turbine systems for providing electrical power into an existing power grid. Subsequently this machine was modified and uprated to 200 kW and re-erected at Clayton, New Mexico, where it generated about 3% of the town's electricity. Three similar machines were erected at other sites in Puerto Rica, Rhode Island and Hawaii.

The next in the series, the MOD-1 windmill became the world's largest machine in May 1979, when it was commissioned. This was also a twin-bladed downwind horizontal axis machine with a blade diameter of 60.96 m and rated at 2 MW. Problems of interference with television signals were overcome but a low level, low frequency noise could only be reduced by lowering the speed of rotation and output. This resulted in design changes in the later machines in the series. The MOD-1 machine was dismantled in 1983[27]. In December 1980 the first of three 2.5 MW MOD-2 machines were commissioned at Goldendale, Washington. These had two blades of 91 m diameter with an upwind orientation and a hub height of 60 m. By grouping three on a fairly small site, it was possible to study the interaction betweenn a windmill cluster. It was orignally hoped to complete testing the three machines by 1983, but various mechanical problems, including cracks in the low speed shafts supporting their rotors, meant that testing would not be completed before the end of 1984[11]. The largest designs in the series, the MOD-5 programme, are based on MOD-2, and also have two blades in the upwind orientation. Rated at 7.3 MW the MOD-5 has a 122 m blade diameter and the MOD-5B 127 m[11]. A MOD-5A was ordered by the Hawaiian Electric Company during 1983 but the order was subsequently cancelled. The reasons for the cancellation were indicative both of the changed political climate in the United States and the short-term effect of excess world oil supplies and lower oil prices. A spokesman for the General Electric Company said that the future market for large wind turbines was very doubtful as forecasts of electricity load growths were lower than expected and subsidies for the use of renewable energy systems in the United States were planned to end[11].

Small to medium range windmills

Multi-bladed windmills for water pumping are still being manufactured in several countries and an estimated one million were in use in the early 1980s[14]. These windmills have a high solidity, or area of blade relative to total swept area. This gives a high starting torque but a relative low power coefficient, typically about 0.2. Wind energy was considered to have a significant role in pumping water in the developing countries by the United Nations Technical Panel[14], but they also identified three problems with existing designs: they were too complicated for local manufacture, too expensive and too difficult to maintain and repair. Several new designs appeared in the late 1970s and early 1980s. These could be made locally and were relatively inexpensive, but a wider educational programme was still needed before the technology could be disseminated[14].

Small low solidity wind turbines for generating electricity in the range up to 10 kW are widely available in many countries. Locally developed windmills in Sri Lanka, for example, can give an output of up to 400 W and would cost no more than 200 US dollars to build[28]. Prototypes are used to charge locally manufactured lead-acid batteries which power low-energy consumption fluorescent tubes. This provides an electric lighting system at about half the cost of conventional kerosene lamps.

Isolated communities in good wind regions, especially in mountain regions, on islands or in coastal areas, can meet their power needs in the 10 to 1000 kW range by a combination of wind power and a suitable back-up system, which could be conventional diesel power or a pumped hydroelectric system[14].

The Vertical Axis Windmill

The modern vertical axis windmill is a synthesis of two earlier inventions. These are the Darrieus[29] windmill with blades of symmetrical aerofoil cross-section bowed outward at their mid-point to form a catenary curve and attached at each end to a vertical rotational axis perpendicular to the wind direction and the Savonius[30] windmill or S-rotor, in which the two arcs of the 'S' are separated and overlap, allowing air to flow through the passage. The Darrieus windmill is the primary power-producing device, but, like other fixed-pitch high-performance systems, is not self-starting. The blades rotate as a result of the high lift from the aerofoil sections, the S-rotor being used primarily to start the action of the Darrieus blades. The wind-energy conversion efficiency of the Darrieus rotor is approximately the same as any good horizontal system[31] but its potential advantages are claimed to be lower fabrication costs and functional simplicity[32]. In 1981 the largest American Darrieus machine, with three blades, had developed 500 kW. A 4 MW machine jointly funded by the Canadian National Research Council and the Institute de Recherche d'Energie du Quebec is due for completion of a site in the St. Lawrence river valley, Quebec, in 1985[11]. An earlier feasibility study concluded that Darrieus machines up to 8 MW in size could be built.

An analysis of the Darrieus rotor suggested to Musgrove[33] that straight-bladed H-shaped rotors, with the central horizontal shaft supporting two hinged vertical blades, could be a more effective system. A variety of designs based on Musgrove's work in the UK during the 1970s and early 1980s have been studied and a small, 6 m diameter, three-bladed version was commercially available by 1980. Musgrove also considered the possibility of siting groups or clusters of windmills in shallow off-shore locations in the UK such as the Wash. Two advantages of this proposal are the higher mean wind speeds and the greatly reduced environmental objections.

During 1983 a new design of vertical axis wind turbine was announced by the Alternative Technology Group of the Open University, UK[11]. The innovative feature is that for any size of turbine the supporting tower need only be 3 to 4 metres high, as the blades are in a 'V' type configuration. A two-bladed version with a fixed tilt angle and bracing wires is illustrated in Figure 9.3. The blades can be hinged for variable tilt angles or rigidly attached to the hub. In high winds the blades can move out to a horizontal position, eliminating the danger of overspeeding, a concept which is also a feature of the Musgrove machines. A single bladed version has also been designed.

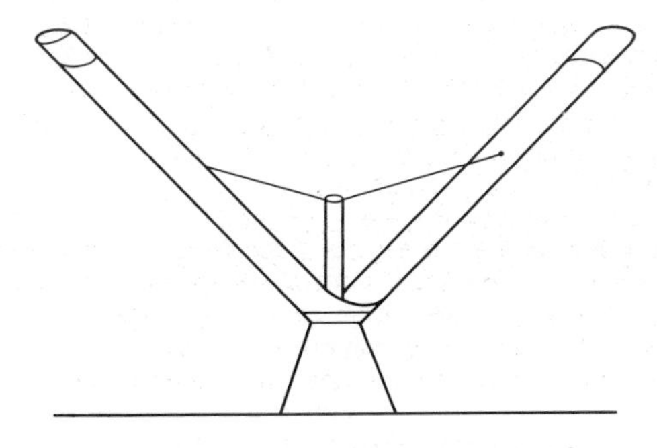

Figure 9.3

Some environmental issues

If a major wind energy programme were to take place in the UK, a recent estimate by the Energy Technology Support Unit suggested that there might be up to 3000 large wind turbines installed[27]. If these were generally dispersed the average distance between machines would be about 7 km. Alternatively some twenty wind farms with areas of 265 km^2 would be needed. Both were not thought to be particularly intrusive. However, visual impact is difficult to quantify and depends on subjective judgements. The blades could also cause a rotating shadow pattern which might present visual problems. A major area of concern has been the danger of birds colliding with the blades, but this was considered to be a negligible hazard[27].

Electromagnetic interference can be caused by a windmill and the considerable amount of research suggests that where significant TV interference appears possible, remedial action should be taken before the problem arises. The area around the Burgar Hill site for the WEG 3MW machine on mainland Orkney had poor reception prior to 1983. A new repeater station was installed so that better signal reception was established before the wind turbines were operating. Noise is also a problem, particularly at low frequencies, and the amount of data is sparse[27]. A large amount of low frequency noise appears to be common to all machines and the only solution is to site the machines sufficiently far away from objectors. This could severely limit potential wind sites. Safety is an obvious problem and blades have been known to become detached and fly off. Some towers have suffered structural failure in high

winds. These potential mechanical design failures can be overcome by good design and sound practice in inspection and monitoring procedures[27].
Many of the problems outlined above could be overcome by the use of the shallow waters round the coast of the UK such as the Wash as possible sites for large arrays. A further advantage is that offshore wind velocities are considerably higher than on many onshore sites.

TIDAL POWER

Tidal mills are known to have been in operation for over eight hundred years and the Romans used tidal currents to power waterwheels in 537 AD[34]. A few of these early installations have survived. Two are on the Rance estuary in Brittany, where the world's largest tidal power plant is sited and another is at Woodbridge in Suffolk[35]. The Woodbridge mill has a basin surface area of some 30 000m² and is filled by the tidal water passing through flap gates in the dam. The water flows out at low tide and drives a 6 metre diameter water wheel which powers four sets of millstones. These early mills operated at very low energy conversion efficiencies and could only be used locally when the tides permitted. Scientific publications on tidal schemes date from the early eighteenth century and various designs for dams and associated turbines appeared from the end of nineteenth century[36]. Modern proposals for exploiting tidal power are based on the use of the stored potential energy in a dam[35]. The use of the kinetic energy of the tidal current has been limited to a few very small-scale developments[37].

Tides are caused by the interaction of the gravitational and kinematic forces of the earth, the moon and the sun. The gravitational force at any point on the surface of the ocean depends on the position of the moon and the sun and on their distance from the point. The period of the tides depend upon the 29.53 day period of rotation of the moon about the earth, the earth's daily rotation and upon the orientation of the earth during its path round the sun[35]. The difference in length between the 24 hour solar day and the 24.813 hour tidal day causes the spring and neap tides. When the sun and moon are almost in line with the earth the tides have their maximum amplitude and are known as the spring tides. When the moon-earth-sun angle is a right angle the tides have their minimum amplitude and are known as the neap tides. The ratio between the greatest spring tide and the smallest neap tide can be up to 3:1[21]. The overall effect of tidal forces is surprisingly small. In the open ocean the tidal range, defined as the difference in amplitude between low and high tides is typically about one metre[21]. Over the continental shelves the tidal range increases to about two metres and in some estuaries or deep narrow bays it can be up to 16 metres. These increased tidal ranges in estuaries or bays are caused by the interaction of two types of wave. The first is the tidal wave advancing from the open sea and the second is the reflected waves from the sides of the estuary[38]. These two waves can reinforce each other at certain times, depending on the shape of the estuary and the period of tide, causing an amplification. Peak amplification occurs at resonance. Theoretically a channel of uniform cross-section would be resonant if its length were equal to one-quarter of the wavelength of the tidal movement[38]. In practice this length is modified by actual variations in depth and width. The rise and fall of the tide is also limited by the frictional losses caused by the action of the water over the sea bed.

Tidal power principles

Tidal power can be obtained from the flow of water caused by the rise and fall of the tides in partially enclosed coastal basins. This energy can be converted into potential energy by enclosing the basins with dams. This creates a difference in water level between the ocean and the basin. The resulting flow of water as the basin is filling or emptying can be used to drive turbo-generators. Electricity conversion removes the geographical restrictions placed on the earlier uses of tidal energy[38]. In Chapter 1 the potential energy of a body of mass m at a height z above the datum line was defined as mgz. If the surface area of a tidal basin is A square metres and the mean tidal range is r metres, then the maximum potential energy available during the emptying or filling of the basin is given by

$$\tfrac{1}{2}\rho g A r^2 \tag{9.5}$$

The tide rises and fall twice during the tidal day of 24.814 hours, so the theoretical average power is four times the maximum potential energy divided by the total line in the tidal day or

$$\frac{4 \times \tfrac{1}{2}\rho g A r^2}{24.813 \times 3600} \tag{9.6}$$

Taking ρ as 1000 kg m^{-3} and g = 9.81 m s^{-2} the theoretical average power becomes

$$0.220\ Ar^2 \tag{9.7}$$

If generation is only on the ebb tide the figure is halved. The actual power output is up to 25% of the theoretical average.
Some locations are particularly favourable for large tidal schemes because of the focusing and concentrating effect which can be obtained from the shape of their bays or estuaries. Typical ranges are shown in Table 9.4. which includes the world's largest tidal range in the Bay of Fundy and Europe's largest, the Severn Estuary.

Table 9.4
Mean tidal range in selected locations(metres)

Location	**range**
Bay of Fundy, Canada	10.8
Severn Estuary, UK	8.8
Rance Estuary, France	8.45
Passamaquoddy Bay, USA	5.46
Solway Firth, UK	5.1

Data from reference 38.

In an ebb generation system the use of pumps to increase the level of water contained in the basin at high tide appears to be attractive. The principle is ilustrated by a simple example. The additional energy required to raise the water level z metres at high tide is

$$\tfrac{1}{2}\ \rho g A z^2 \tag{9.8}$$

The maximum potential energy now available during the emptying of the basin becomes

$$\tfrac{1}{2}\ \rho g A (r+z)^2 \tag{9.9}$$

giving a net gain compared with equation 9.4. of

$$\tfrac{1}{2}\ \rho g A (r^2 + 2rz + z^2 - z^2 - r^2)$$

$$= \quad \rho g A r z \tag{9.10}$$

In practice there are two problems. Pumping involves some loss of overall efficiency and turbines capable of pumping as well as generating are more expensive. The power needed for pumping may be required when demand on the whole electricity system is high and could involve the use of an expensive form of generation in another section of the network[21].

Tidal power schemes

There are a number of different schemes which can be grouped into two main combinations, depending on whether one or two basins are used:

(i) Single basin, generation only on the ebb tide

(ii) Single basin, generation only of the flood tide

(iii) Single basin, generation with both the ebb and the flood tides

(iv) Single basin, generation with both tides and pumped storage

(v) Double basin

(vi) Double basin with a pumped storage system.

Any particular scheme could be optimized against any one of a number of different and distinct parameters. These include maximum net energy output; constant power output; constant head operation; maximum pumped storage capacity or lowest initial capital requirements[38].

Single basin ebb generation allows the incoming tide to flow through sluice gates and the turbine passageways. These are closed at high tide and the water is retained until the sea has ebbed sufficiently for the turbines to operate. This is normally at about half the tidal range. Initially the flow is restricted to maintain a high head and to operate the turbines at maximum efficiency. Later in the cycle the turbines are usually operated at maximum power[21].

Single basin flood generation is the reverse of ebb generation. It has a number of potential disadvantages, the main one being the prolonged periods of low-tide experienced above the dam. A second disadvantage is that the amount of energy would be less than with an ebb generation scheme, as the surface area of the estuary decreases with depth.

Two-way generation with a single basin system generates electricity from both the flood and ebb tide. This does not result in a greatly increased power output. Neither phase of the cycle can be taken to completion because of the need to reduce or increase levels in the basin for the next phase. There are also economic disadvantages. The turbines are more complex and less efficient if they are required to operate in both directions and the turbine water passages must be longer. An advantage is that power is available four times in the tidal day, rather than for two longer periods.

Double-basin schemes often include provision for pumped storage, but in their simplest form they could operate as two independent two-way generation schemes. In another form water would always flow from the higher level basin to the second lower level basin. The second basin could only be emptied at low tide.

A detailed discussion of the relative merits of the different schemes has been given by Taylor[21], who points out that it is difficult to generalise as a large number of variables, which vary from one site to another, need to be considered.

Tidal power sites

Many of the various design studies which were carried out until the early 1970s suggested that while tidal power systems were technically possible they would be unable to generate electricity at a competitive price. A notable exception was the first major report on the Severn Barrage, published in 1933[39], but the recommendations were ignored and over half a century later further feasibility studies were still being carried out[11]. Only three modern tidal power schemes were operating in 1984. The largest and oldest is the Rance Barrage near St. Malo on the Brittany coast of France. Two much smaller schemes are in the USSR and China. All three schemes have been built primarily to gain operating experience for the possible development of much larger systems[21]. Work on the Rance site commenced in June 1960, the final closure of the estuary against the sea took place in July 1963 and the last of the twenty four 10 MW turbo generators was commissioned in November 1967[35]. The overall width of the Barrage is 750 m. The tides follow a fairly constant two week cycle throughout the year. During the first week of the cycle the tidal range is between 9 m and 12 m and in the following week between 5 m and 9 m[35]. The mean tidal range is 8.45 m. For the lower tidal ranges the barrage operates only on the ebb tide with the basin level increased by pumping. For mean and spring tides two way generation is used, sometimes augmented by pumping. Electricite de France have shown[35] that the output/input ratio for pumping can be as high as 2.8:1. The operation of La Rance is computer controlled and optimized to match the period when it would be most expensive to generate electricity for the French national grid from conventional power stations. The nominal average output of between 50 and 65 MW is therefore not the maximum which could be obtained.

In the USSR a small 400 kW pilot scheme was completed in 1968 at Kislogubsk on the Barents Sea. The main objective was to try out a new construction method, the use of floated-in prefabricated caissons to form both the main powerhouse and spillway structures[40]. The overall dimensions of this structure were 36m x 18.3m by 15.35 metres high and the single reversible turbine was purchased from the French company that supplied turbines for the Rance Barrage[35]. The People's Republic of China have a broadly similar pilot scheme rated at 500 kW at Jangxia Creek in the East China Sea[21].

Potential tidal sites

In the UK one site has been outstanding for nearly seventy years, the River Severn. Some of the wider issues which are mentioned towards the end of this section could also be considered in relation to some other potential sites. Government interest in the Severn commenced in 1925 when the House of Commons established a Sub-Committee, which later became the Severn Barrage Committee of the Economic Advisory Council. The main conclusion of its report in 1933 was that the cost of power generated by a Severn Barrage with secondary storage some 20 km away would be only two-thirds of the cost of that generated at equivalent coal-fired stations[39]. The scheme included road and rail crossings and the rated output was 804 MW. A further report in 1944 suggested doubling the output of the turbines to 25 MW while maintaining the rated output at some 800 MW. A later report in the early 1950s drew attention to the potentially high capital cost of any scheme. Subsequent studies included the single basin, ebb generation 'Wilson Scheme' proposed by E.M. Wilson in 1968[41], in which a 14.6 km barrage would be built between Lavernock Point on the Welsh coast to Sand Point, a few kilometres north of Weston-super-Mare; the two basin schemes proposed by T.L. Shaw[42] and the Central Electricity Generating Board[43]; and a system of bunded reservoirs[44]. Published estimates of the power that could be produced from the various schemes up to 1975 ranged from 8 to 42 TWh[43] and in a comment on the various proposals that had been made over the past fifty years, the authors of an Institution of Chemical Engineers report[45] stated in 1976

> A curious feature has been the regular conclusion that the scheme would have been economic if embarked upon on earlier occasions, but never on the current one.....

In 1977 the Department of Energy summarized the various proposals[38] including the results of their own specially commissioned studies[46,47]. They noted that these had produced conflicting results in assessing one of the most basic parameters - the effect of the barrage on the tidal range. NEDECO[47] predicted a decrease of 1 m while the UK Hydraulics Research Station[46], using a more complex mathematical model predicted an increase of about 1.4 m. This was followed in 1978 by the establishment of a further pre-feasibility Severn Barrage Committee which reported in 1981[48], that it is

> technically feasible to enclose the estuary by a barrage located in any position east of a line drawn from Porlock due north to the Welsh coast

The most cost-effective of three schemes considered in detail was a single basin, ebb-generation scheme with a 13 km barrage from Brean Down, a few kilometres south of Weston-super-Mare, to Lavernock Point, with an estimated annual output of 13 TWh at a cost of some 2.4 pence per kWh, at that time close to the official cost of nuclear electricity. This particular scheme is the subject of a two year study jointly funded by Government and Industry which commenced in June 1983[11]. While this study was in progress, a rival scheme for a smaller 6.7 km barrage between Severn Beach and Sudbrook Point in Gwent was proposed early in 1984 by a private consortium headed by the UK building group Wimpey[11]. This would have a proposed annual output of 2.43 TWh, but would have the added advantage of providing a second road crossing near the existing Severn Bridge. Other possible smaller sites in the UK include Morecambe Bay, the Humber, the Wash and the Solway Firth.

Canada. Tidal ranges of up to 16 metres have been recorded in the upper regions of the Bay of Fundy in North East Canada. This has been the subject of several investigations including the working of the Atlantic Tidal Power Board[49] in 1969 (development not economically justified) and the Bay of Fundy Tidal Power Review Board[50] in 1977 who assessed the potential of thirty possible sites. The three with the best prospects, Cobequid Bay, Shepody Bay and Cumberland Basin, had an estimated potential output of 6.4 GWe. A 17.8 MW turbine generator with an Escher Wyss-Stratflo rim-mounted generator was selected for installation in an existing barrage across the Annapolis River at Annapolis Royal, Nova Scotia, in the early 1980s[21,50]. This type of compact turbine can be built without a central shaft, as the bearings are hydrostatic pads on the outer rim, and could result in much lower barrage construction costs.

People's Republic of China. Estimates of some 500 possible sites have suggested a potential of over 110 GW[21].

USSR. Estimates have concentrated on the White Sea between Murmansk and Archangel where the potential could range from 16 to over 50 GW.

Other parts of the world with potential large-scale sites include the Kimberley Region of Western Australia, several countries in South America, India and Korea.

A series of estimates reviewed by Hubbert[51] suggested that the total tidal energy dissipated in the world's shallow seas was no more than 10^{12} watts although this data appeared to omit any estimates from the People's Republic of China. The average maximum potential power which could be recoverable from these sites was estimated to be 64 GWe in 1969. When it is appreciated that some of the sites are very long distances away from any potential main user, this figure does not seem to be unduly pessimistic.

Possible impacts of tidal schemes

It is difficult to quantify the social, industrial and environmental impacts which any proposed scheme in the UK or elsewhere could have. These have been widely reviewed in the literature[21,52] and some of the main points are discussed briefly below.

Water levels both in the basin upstream of the barrage and to seaward could be changed.
Tidal flows reduce the strength of the currents upstream of the barrage. Downstream and to sea the effects could extend over 50 km.
Sedimentation may occur in the basin and could lead to a slow and possibly small reduction in basin volume. To seaward the sediments previously swept out by tidal flows may stay deposited.
Mixing will occur less in the water above the basin because of reduced currents and tidal excursions.
Navigation. Ships could be slowed by passing through locks, on the other hand, predictable periods of deeper water could be an advantage.
Industry could benefit during construction but may have to adopt higher standards in dealing with possible polluting liquid effluents.
Land Drainage could be affected inside the barrage because of higher low water levels.
Sea Defences will be less liable to storm damage after the construction of a barrage.
Ecosystem. The acquatic ecosystem will always be affected by any changes in turbidity and salinity.
Migratory Fish will face the obstacle of the barrage, but the inward journey will probably be straightforward through the sluices in ebb generation schemes.
Recreational opportunities could be enhanced in suitable locations, with less turbid water above the barrage.

WAVE POWER

The history of wave energy conversion probably dates from the last few months of the eighteenth century, when the first patent for a wave energy device was filed in Paris[54]. This early concept envisaged a giant lever, with its fulcrum on the shore, attached to a floating body which rose and fell with the waves. Since then there have been hundreds of patents filed throughout the world with well over three hundred in the UK alone between 1856 and 1973[55]. Salter[56] comments that there was

> a spate of them after the First World War. Flaps, floats, ramps, converging channels and liquid pistons are all advocated, and much ingenuity is expended in accomodating tidal rise and fall....

Only a few proposals appear to have had model tests and these showed poor efficiency. Modern developments started in 1945 when Yoshio Masuda commenced privately funded research in Japan on a wide range of devices[57]. In 1960 his work received government support and he concentrated on wave-activated air turbines, one of which was installed in a lighthouse in Tokyo Bay. This had a maximum output of 130 watts[57]. By the early 1970s over two hundred small units were operating in Japan, mostly on buoys. The UK Department of Energy wave energy programme commenced in 1974. This early work showed that the waves arriving at Britain's Atlantic coastline delivered a surprisingly large amount of energy[43,58] and the national programme quickly established that it was physically possible to generate useful power from ocean waves with reasonable efficiency[59]. However, after a decade of development, always exciting and sometimes frustrating, none of the devices investigated seemed to be capable of generating power at a cost which would be comparable to more conventional sources[59]. Some of the more promising concepts are reviewed briefly after the introduction to simple wave theory in the next section.

Wave theory and characteristics

The basic characteristics of wave power can be studied from standard linear wave theory[60]. This considers a simple progressive sinusoidal wave of amplitude a, wave length λ and period T, progressing in deep water. Sinusoidal waves of a single wavelength are known as monochromatic waves and 'deep water' is defined as having a depth greater than half a wavelength. The velocity with which the wave propagates, the phase velocity, can be written as

$$\frac{\lambda}{T} \quad \text{or} \quad \frac{gT}{2\pi} \tag{9.11}$$

By considering the rate of change of potential energy as the water in a wave above sea level falls into the troughs in front of the wave, the power in a wave front of width W is given by[56]

$$\frac{W\rho g^2 a^2 T}{8\pi} \tag{9.12}$$

For example a monochronatic wave in water with a density of 1.026 kgm^{-3}, a 1 metre amplitude and a period of 15 seconds would give a power output of 58.93 kW per metre width of wave front. The wavelength and period are simply related, as shown in equation 9.11, so that the wavelength in this case is 351 m and the velocity 23.4 ms^{-1}.

Real ocean waves are quite different from the theoretical ideal wave described above. Ocean waves are generated by the wind, so that wave energy is an indirect form of solar energy. The ocean acts effectively as an extremely large integrator for wind energy[21,61] and in addition the inertia of the water can provide a limited amount of short-term energy storage which can compensate for variations in wind velocity with time and place[43]. The waves arriving at a point can have originated from storms hundreds of kilometres away, the 'swell' sea, or from local winds, the 'wind' sea[61]. The distance from the origin of swell waves is known as the fetch. Swell waves can appear to be substantially plane and monochromatic so that from equation 9.11, the longer the period between waves the faster they travel. However, the local wind sea which is superimposed on it can be more complicated and random in wavelength, phase and direction. Any record of sea waves is therefore very complex and is best described as 'the linear sum of many monochromatic waves of random relative phase distributed both in direction and across the frequency spectrum'[61]. A detailed discussion of the spectral density function has been given by Pierson and Moskowitz[62]. However, for most practical purposes a very simplified relationship has been derived which is probably accurate to within ± 30%. This gives the power of a wind generated wave system for any location in terms of the significant wave height H_S, defined as the average height of the highest third of the waves, and the zero crossing period T_Z, defined as the time interval between successive upward movements of the water level past the mean position. For the Ocean Weather Ship India (50°N, 19°W) the power per metre of wave front has been shown[63] to be approximately

$$0.55\, H_S^{\,2} T_Z \text{ kW} \tag{9.13}$$

The estimates of power availability for wave energy systems in the mid 1970s were based on India data which suggested that an average power of 91 kW per metre of wave front was available[58]. This ranged from periods with very little power to severe storm conditions when the power level could exceed several Megawatts[64]. More recent measurements near South Uist reported by the CEGB in 1983 showed that at inshore sites more suitable for the deployment of wave energy systems, power levels between 40 and 50 kWm^{-1} could be expected in water about 50 metres deep. Levels of 25 kWm^{-1} could be expected off the north-east coasts of England or the south-west of Wales.

Types of wave energy convertor

There are been a number of different classifications in the past decade, starting with Salter's flaps, floats, ramps, converging channels and liquid pistons or air bells[56]. The concept of 'active' and 'passive' systems has also been suggested[65], active systems having parts which respond to the waves with power generated from the relative motion of these components while passive systems absorb wave energy by virtue of a fixed structure. One of the simplest methods for using wave energy would be to construct an immovable structure to capture large volumes of water which can subsequently be used to drive a water turbine. This system could also be classified as a ramp or passive system. The best known proposal was initiated in Mauritius in the 1950s and has been the subject of a series of official reports, summarized by Bott[66] in 1979. All the reports agreed that the project would be technically feasible, but the economic viability has proved to be the stumbling block. By the early 1980s the more generally accepted scientific classification was that of terminators, attenuators and point absorbers[64].

A terminator is defined as a wide structure which is aligned perpendicular to the incident wave direction. Much of the experimental work on this type has been carried out in narrow wave tanks and thus has resulted in good agreement with extensive theoretical studies so that the performance of terminators is generally more fully understood than other types of device. The best known is probably the 'Salter Duck', a system originally proposed by Salter in 1974[56]. The floating 'duck' section is an asymmetric cam-shaped device designed to extract energy through semirotary motion induced by the incident waves[64]. The system would consist of a long central core, or spine, upon which the duck sections are mounted. Large gyroscopes would be placed inside the nose of each duck, which rotates along its principal axis. The precessing motion of the gyroscopes could be used to drive hydraulic pumps. The concept is revolutionary, as it would be designed to be maintenance free over a 25 year lifetime. The Cockerell raft was based on a simpler concept, that of a series of pontoons or rafts connected by hinges, with power generated from the relative movement of the rafts. Both the Duck and the Raft were taken to 1/10 scale model tests[67,68]. Several systems known as oscillating water column devices and which could also be considered as variations of Masuda's original air bell concept have been studied. One which might proceed to full-scale testing is the National Engineering Laboratory's breakwater system, which consists of a concrete structure mounted on the seabed. The motion of the waves causes a column of water inside the structure to rise and fall, inducing air flow through turbines. Proposals were suggested in 1982 by a private consortium and the National Engineering Laboratory for 4 MW prototype to be built off the island of Lewis in the Outer Hebrides but no decision had been reached two years later[11]. Other terminators have been studied in the UK, including the Russell Rectifier and the Sea Clam and detailed descriptions are widely available[21].

An attenuator is a long thin structure which is aligned parallel to the incident wave direction. It was originally thought that energy could be progressively extracted along its entire length, but this has proved impractical because the rear element would need to extract as much energy as the front one for optimal operation[64]. Two devices considered in the UK were the Vickers Attenuator and the Lancaster flexible bag. Neither device was able to show a satisfactory performance in model testing. In 1978 the Japanese built a special floating test bed far oscillating water column devices, the Kaimei. It was 80 m long, 12 m wide and 7.5 m deep, contained 22 individual water compartments and had a rated capacity of 2 MWe. Model tests by the CEGB showed that it would have a very poor performance[64] and its actual measured output was less than 2% of its anticipated average output[69]. It was clear that the Kaimei had not been optimized.

The point absorber is an axially symmetric device, constrained to move vertically, which can absorb wave energy from any direction. Theoretically it can absorb wave energy from an effective wave frontage of $\lambda/2\pi$[64]. This means that a number of interconnected but widely distributed point absorbers could produce as much power as a continuous line absorber having the same total length. Research into the development of composite materials at Queen's University Belfast resulted in the development of a glassfibre reinforced polyester resin and its utilization in the prototype of a new type of wave energy convertor, the Belfast Buoy. It can be considered as an oscillating water column device with an important difference, the vertical axis air turbine rotates in the same direction irrespective of the direction of air flow. The turbine system has been named after its inventor, Professor Alan Wells[70]. Several different configurations have been suggested, but a good hydrodynamic performance can only be obtained from a very limited bandwidth[64].

Summary

Unlike the other alternative energy sources geothermal energy is capable of providing continuous heat and power. With electricity generation, the plant is particularly suitable for base load operation. The use of low-grade aquifers can provide space heating at costs comparable or below conventional systems. The long term future is fascinating. Several authorities believe that heat mining, the exploitation of hot, dry rocks, could become a commercial reality within the next two decades[1,7]. Should this happen, a major new energy resource, comparable in size to the ultimate oil and gas resources, would be available.

Wind energy systems are already widely distributed throughout the world for small-scale pumping and mechanical power applications. Recent developments in small to medium scale windmills could greatly increase their use, particularly in the developing countries. By 1984 the economics of the large-scale wind turbines in the 3 to 7 MW range appeared to be attractive, although several years of testing were still needed.

The contribution which tidal energy could make to the total world energy demand is negligible, as there are very few potentially suitable sites. For those countries where tidal energy could be exploited, it could provide a small but predictable and highly reliable source of electricity[40]. In the UK the Severn scheme has been studied very extensively for over fifty years, but the latest proposals indicate that probably no more than 5% of the UK electricity demand could be met.

The future of wave power is particularly uncertain. None of the systems tested in the past decade have been able to demonstrate that they could generate electricity at a cost which would be comparable with other, more conventional sources, except for a few specific locations in remote islands. The economics would alter in favour of wave power if conventional methods become more expensive, which many authorities believe is inevitable.

References

1. Armstead, H.C.H., Geothermal Energy, Second Edition, E. and F.N. Spon Ltd, London and New York, 1983.

2. Holland, M.B. Power from the Earth, Chartered Mechanical Engineer, 40-45, November 1978.

3. Dunham, K., Geothermal Energy and Heat Pumps, Keynote Address, Int. Conf. Future Energy Concepts, Institution of Electrical Engineers, London, 1979.

4. Shaw, J.R. and Robinson, P.E., Prospects for the future exploitation of geothermal energy, Int. Conf. Future Energy Concepts, Institution of Electrical Engineers, London 1981.

5. Pearson, C.M., Exploitation of Hot Dry Rock Geothermal Energy in the United Kingdom, Int. J. Ambient Energy, 3,1, 19-26, 1982.

6. Hubbert, M. King, The Energy Resources of the Earth, Scientific American, 225, 61-70, September 1971.

7. Leardine, T., Geothermal Power, Phil Trans. R. Soc. Lond. A., 276, 507-526, 1974.

8. World Energy Resources: 1985-2020, World Energy Conference, London, 1980.

9. Garnish, J.D., Geothermal energy and the UK environment, in Energy and our Future Environment, Institute of Energy, London, 1983.

10. Report of the Technical Panel on Geothermal Energy (second session), United Nations Conference on New and Renewable Sources of Energy, Nairobi, 1981.

11. World Solar Markets, Financial Times Business Information Ltd., London, 1983 and 1984.

12. Garnish, J.D., Prospects for Geothermal Energy, Keynote Address, Int. Conf. Future Energy Concepts, Institution of Electrical Engineers, London, 1981.

13. McVeigh, J.C., Sun Power, Second Edition, Pergamon Press, 1983.

14. Report of the Technical Panel on Wind Energy (second session), United Nations, Conference on New and Renewable Sources of Energy, Nairobi, 1981.

15. Juul, J., Wind Machines, Wind and Solar Energy Conference, New Delhi, UNESCO, 1956.

16. Golding, E.W. and Stodhart, A.H., The use of wind power in Denmark, ERA Technical Report C/T 112, 1954.

17. Gimpel, G., The windmill today, ERA Technical Report IB/T22, 1958.

18. Putnam, P.C. Power from the Wind, Van Nostrand, New York, 1948, and Second Edition with G.W. Koeppl, 1982.

19. Golding, E.W., The generation of electricity by wind power, E. & F. Spon, 1955. Reprinted CTT 1976.

20. Golding, E.W. and Stodhart, A.H., The potentialities of windpower for electricity generation, British Electrical and Allied Industries Research Association, Tech. Rep. W/T16, 1949.

21. Taylor, R.H., Alternative Energy Sources, Adam Hilger Ltd., Bristol, 1983.

22. Davenport, A.G., Proceedings of the (1963) Conference on Wind Effects on Building and Structure Vol. 1, HMSO, 1965.

23. Caton, P.G., Standardised maps of hourly mean wind speed over the United Kingdom and some implications regarding wind speed profiles, Fourth International Conference on Wind Effects on Building and Structures, London, 1975.

24. Hinrichsen, D. and Cawood, P., Fresh breeze for Denmark's windmills, New Scientist, 567-570, 10 June 1976.

25. Lindley, D. and Stevenson, W., The horizontal axis wind turbine project on Orkney, Proc. Cranfield, Third BWEA Wind Energy Conference, 16-23, 1981.

26. Reed, J.W., Maydew, R.C. and Blackwell, B.F., Wind energy potential in New Mexico, SAND-74-0077, Sandia Laboratories Energy Report, July 1974.

27. Bedford, L.A.W. and Tolland, H.G., Wind energy and the environment, in Energy and our Future Environment, Institute of Energy, London, 1983.

28. McVeigh, J.C., When nuclear power is not the answer, Electrical Review International, 1, 1, February 1984.

29. Darrieus, G.J.M., Turbine having its rotating shaft transverse to the flow of the current, US Patent 1,835,018, 8 December 1931.

30. Klemin, A., The Savonius wing rotor, Mechanical Engineering 47, No. 11, November 1925.

31. South, P. and Rangi, R.S., A wind-tunnel investigation of a 14 ft diameter vertical-axis windmill, National Research Council of Canada, LTR-LA-105, September 1972.

32. South, P. and Rangi, R.S., The performance and economics of the vertical-axis wind turbine developed at the National Research Council, Ottawa, Canada, Agricultural Engineer, February 1974.

33. Musgrove, P.J. and Mays, I.D., 'The Variable Geometry Vertical Axis Windmill', Proc. 2nd Int. Symposium on wind energy systems, BHRA Fluid Engineering, Cranfield, E4, 39-60, 1978.

34. A History of Technology, Oxford University Press, 11, 593-611, 1956.

35. Holland, M.B., Power from the Tides, Chartered Mechanical Engineer, 33-39, July 1978.

36. Charlier, R.H., Tidal power plants: sites, history and geographical distribution, Proc. 1st Int. Symp. on Wave and Tidal Energy, BHRA, Cranfield, 1979.

37. Wyman, P.R. and Peachey, C.J., Tidal Current Energy Conversion, Int. Conf. Future Energy Concepts, 164-169, Institution of Electrical Engineers, London, 1979.

38. Tidal power barrages in the Severn Estuary, Energy Paper Number 23, Department of Energy, HMSO, London, 1977.

39. Severn Barrage Committee Report, 1933, quoted in The Exploitation of Tidal Power in the Severn Estuary, Fourth Report from the Select Committee on Science and Technology, HMSO, London, 1977.

40. Tanner, R., Murphy, D and Warnock, J.G., Influence of Technological Advances on Potential Tidal Power Developments, Int. Conf. Future Energy Concepts, 130-142, Institution of Electrical Engineers, London, 1979.

41. Wilson, E.M., Severn, B, Swales, M.C. and Henery, D., The Bristol Channel Barrage Project, Proc. Eleventh Conf. on Coastal Engineering (1968), ASME, New York, 1969.

42. Shaw, T.L., Tidal Power, Journal of the Royal Society of Arts, 726-8, November 1976.

43. Denton, J.D. et. al., The potential of natural energy resources, CEGB Research, 2, 36-39, May 1975.

44. Russell, R.C. in Severn Barrage Seminar, Energy Paper Number 27, Department of Energy, HMSO, London 1977.

45. Materials and Energy Resources, The Institution of Chemical Engineers, London, 1976.

46. Severn Barrage Study, Ex 753, Hydraulics Research Station, Wallingford, November, 1976.

47. Pre-feasibility study on the closure of the Estuary, Report for the UK Department of Energy, Netherlands Engineering Consultants, March 1977.

48. Tidal power from the Severn Estuary, Energy Paper Number 46 (two volumes), Department of Energy, HMSO, London, 1981.

49. Feasibility of Tidal Power Development in the Bay of Fundy, Atlantic Tidal Power Programming Board, Ottawa, 1969.

50. Reassessment of Fundy Tidal Power, Bay of Fundy Tidal Power Review Board, Ottawa, 1977.

51. Miller, H., Strub, W. and Wilson, E.M., Optimising Stratflo turbines for tidal applications, Proc. 2nd Int. Symp. on Wave and Tidal Energy, BHRA, Cranfield, 1981.

52. Hubbert, M. King, Energy Resources, in Resources and Man, National Academy of Sciences, W.H. Freeman and Company, San Francisco, 1969.

53. Ruxton, T.D., Tidal power from the Severn Estuary: an initial environmental impact assessment, in Energy and our Future Environment, Institute of Energy, London, 1983.

54. Ross, David, Energy from the Waves, Second Edition, Pergamon Press, 1982.

55. Holland, M.B., Power from the waves, Chartered Mechanical Engineer, 41-4, September, 1978.

56. Salter, S.H., Wave Power, Nature 249, 5459, 720-724, June 21, 1974.

57. Masuda, Yoshio, Study of wave activated generator and future view as island power source, Second Int. Ocean Development Conference, 2074-2090, 1973.

58. Mollinon, D., Buneman, O.P. and Salter, S.H., Wave power availability in the NE Atlantic, Nature 263, 5574, 223-226, September 16, 1976.

59. Andrews, S.A., and Platts, M.J., Wave energy and the environment, in Energy and our Future Environment, Institute of Energy, London, 1983.

60. Milne-Thompson, L.M., Theoretical Hydrodynamics, Fourth Edition, Macmillan Press, London, 1962, or in Coulson, C.A., Waves, Oliver and Boyd, Edinburgh and London, 1955.

61. Glendenning, I., The potential of wave power, Applied Energy, 3, 197-222, 1977.

62. Pierson, W.K. and Moskowitz, L., A proposed spectral form for fully developed wind seas, J. Geophysical Res., 69, 24, 1964.

63. Count, B.M. and Robinson, A.C., On the estimation of power from limited data, CEGB Report R/M/N 895, London, 1976.

64. Count, B.M., Fry, R. and Haskell, J.H., Wave Power: the story so far, CEGB Research, 15, 13-24, November 1983.

65. Glendenning, I. and Count, B.M., Wave Power, Proc. Symp. Renewable Sources of Energy, 50-81, Royal Society of Arts, London, June 1976.

66. Bott, A.N. Walton, Electro/Mechanical aspects of the Mauritius 'passive' type wave energy project, Int. Conf. Future Energy Concepts, 81-87, Institution of Electrical Engineers, London, 1979.

67. Platts, M.J., The development of the wave contouring raft, Ibid, 160-166.

68. Bellamy, N.W., Wave Power experiments at Loch Ness, Ibid, 167-176.

69. Salter, S.H., Private communication, 1983.

70. McIlhagger, D.S. and Long, A.E., Harnessing Energy from the waves, Queen's University Association Annual Review, 48-52, 1979.

Further Reading

Alternative Energy Sources by Dr. R.H. Taylor, Adam Hilger Ltd., Bristol, 1983, covers all the topics in this chapter in greater breadth and depth. Taylor has written for the engineer or scientist who wishes to be introduced to the subject and his treatment reflects a more mathematical approach. Geothermal Energy by H.C.H. Armstead, E. and F.N. Spon Ltd, London and New York was published in a revised and updated version in 1983 and gives an authoritative overview of all aspects of geothermal energy. Geothermal Energy Projects, edited by L.J. Goodman and R.N. Love, Pergamon Press, 1980, includes a detailed examination of New Zealand, Hawaii, the United States and the Philippines. Among many books on windpower, Dr Gerald Keopp's Putman's Power from the Wind, Van Nostrand Reinhold Company, 1982 not only covers the history of the first megawatt scale wind turbine, but also introduces many of the later large machines. Tidal power is specific to particular sites, so for the UK the Department of Energy's Tidal Power from the Severn Estuary, Energy Paper Number 46, in two volumes, London, HMSO, 1981, is a very useful publication. A broad, non-technical overview of the historical background and development of wave power is given by David Ross in Energy from the Waves, Second Edition, Pergamon Press 1982.

Exercises

1. Water for a geothermal district space heating project leaves the well at 76°C at a flow rate of 20 litres per second. It is used to supply heat above a datum temperature of 40°C. If the heating season lasts for 173 days, how much oil does the project save if the overall combustion efficiency of the oil burner is 75%?

2. According to Garnish[9], geothermal extraction should be considered as mining the earth's heat. How did he reach this conclusion? What progress has been made since 1983 in the hot, dry rocks project?

3. One possible source of geothermal energy is the geopressurized fluids associated with sedimentary rocks. What recent action has been taken to assess the potential of this source?

4. The following data relates to three wind turbines described in the text. Determine an overall power coefficient for each machine. What conclusions can you draw from this? Take ρ for air as 1.293 kg m^{-3}

	Rated Power MW	**Rated speed** ms^{-1}	**Blade diameter** m
WTS-4	4.0	13.7	78
WEG Orkney	3.0	17.0	60
MOD-2	2.5	12.3	92

5. The Central Electricity Generating Board are assessing the performance of several wind turbines. What is the latest information on their tests? If these tests prove satisfactory, what percentage of the UK electricity demand could be met by wind turbines in 25 years time?

6. What are the main differences between the horizontal and vertical axis machines? How does the performance of the two largest machines of each type compare?

7. The tidal basin at Rance has an area of 22 square kilometres and the mean tidal range is 8.45 m. The averaged mean output from the station in 1977 was 52 MW. What percentage of the theoretically available power does this represent?

8. 'The amount of energy potentially available in a single basin flood generation scheme would be less than with an ebb generation scheme as the surface area of the estuary decreases with depth'. Show that this statement is correct by considering an estuary with a V-shaped vertical cross-section.

9. Explain how the tidal day has been estimated to be about 48 minutes 46 seconds longer than the 24 hour solar day.

10. State whether the following statements are true or false:

    (a) With a monochromatic wave the velocity of propagation increases if the period T increases.
    (b) The significant wave height is the average height of the ocean waves at a particular location.
    (c) Wave energy has a very wide range of wave components, each of which has its own period and wave height.
    (d) The Japanese had a wave energy device delivering 2 MWe to their national grid system in 1980.

(e) Wave energy devices have been in use for at least a quarter of a century and are very economic for certain applications.
(f) A terminator is aligned parallel to the wave crests and forms a barrier to the oncoming waves.

11. The power output per metre width of monochromatic wave front is 22.1 kW. The water has a density of 1.026 kgm^{-3} and the wave period is ten seconds. Show that the wavelength is about 156 m and that the maximum trough to crest height is 1.5 metres.

12. Trace the history and development of the oscillating water column or air bell systems. What are the latest design proposals from the UK National Engineering Laboratory?

Answers

1. 1400 tonnes (using standard heating value)

4.

WTS-4	0.4
WEG	0.262
MOD-2	0.246

The projected performance of WTS-4 looks optimistic.

7. 15%

10. (a) True
(b) False
(c) True
(d) False
(e) True
(f) True

10. WHAT NEXT?

The previous chapters have examined trends in population growth, gross national product and in the growth of the production and consumption of the major energy resources. It has been shown that the use of simple exponential growth models to describe these trends can lead to errors. The use of logistic equations and other limited growth mathematical models in the life sciences was established in the nineteenth century, but it is only in the past two decades that their application to energy resources has been appreciated as a result of Hubbert's work[1]. There is really no immediate concern about possible shortages in any of the world's major energy resources provided growth rates remain less than some 2% per annum. However, there could be very serious problems if growth rates in production and consumption started to increase again. This is shown in Table 10.1, where the estimated life of the proven reserves of three fossil fuels, A, B and C is shown for a number of different growth rates. With a constant annual production rate, their lives would be 30, 50 and 200 years respectively. A one per cent per annum growth rate in consumption reduces the life of fuel A by just over a tenth, but nearly halves the life of fuel C. Growth rates approaching 10% per annum are clearly unsustainable and would make nonsense of long term energy planning.

Table 10.1

Lifetime (in years) of three fossil fuels at different growth rates

Growth rate%	A	B	C
nil	30	50	200
1	26	41	110
2	24	35	81
3	22	31	66
5	19	26	49
10	15	19	32

By the end of 1983 the combined proven reserves of oil and natural gas were at least forty times greater than annual production and consumption rates. Coal resources were known to be at least five times greater than all known or possible oil and natural gas resources. Hydropower and nuclear energy provided about one tenth of the world primary energy consumption, but their contribution could very probably increase from the 1983 level of some 700 million tonnes of oil equivalent to over 1500 million tonnes by the end of the century when existing and planned developments had been completed.

Some months before the events of October 1973 Tiratsoo[2] quite correctly predicted the problem facing the world. It was 'not so much a forthcoming oil shortage but rather the fact that the era of cheap oil on which our modern civilisation has largely been built has now come to an end'. This has caused economic, political and organizational problems for all countries since then. The poorer developing countries who are unable to meet their own requirements for commercial energy from their indigenous resources have been caught in what has been described as an 'energy trap'[3]. They need more energy for development but two world recessions in the past decade have reduced the demand for their exported products. The revenue from these products has fallen and they have not been able to afford the increased prices for oil. Many developing countries are also facing another energy crisis, acute shortages of fuel wood caused by the rapid and massive destruction of their forests[3,4,5]. The Centrally Planned Economies have been able to keep energy supply and demand in balance while the Industrial Economies have been able to reduce their oil dependence as shown in Chapter 2, mainly through energy conservation and fuel substitution.

In parallel with these major readjustments to energy consumption and production patterns another philosophy, that of the Conserver Society, was beginning to attract attention, particularly in Canada[6], the country with the highest per capita energy consumption in the world in 1982. The Science Council of Canada was not only concerned about the limitations of material resources but also on how those limitations should cause the industrial societies to rethink their basic priorities. Resource constraints made patterns of economic development which relied on the continuous expansion of material production unsustainable. These patterns were also seen to have progressively less to do with actually making life better, easier or more rewarding[7]. The reasons behind these arguments are discussed in the following section.

Economics, Energy and Entropy

One statement of the Second Law of Thermodynamics, or the Entropy Law, is that heat flows by itself only from the hotter to the colder body without the addition of external work. Another statement is that the entropy of a closed system (one which exchanges only energy with its surroundings) continuously increases towards a maximum, as shown in Chapter 1. Professor Nicholas Georgescu-Roegan[8], the distinguished Romanian economist, extended Professor Soddy's critisisms of traditional economic theories[9] by showing that the amount of low entropy energy and material available for use by man is gradually decreasing as both energy and matter are dissipated to the high entropy state by the economic process. By continuously depleting our stocks of low entropy (for example the fossil fuels) then our materials and energy resources will be denied to future generations. The traditional 'market' economic theories cannot reconcile the values which different generations assign to resources. He pointed out that it is quite unsound to think that the economic system will automatically adjust itself through the price mechanism. More production means an increasing amount of waste and pollution though the conversion of low entropy raw materials into high entropy goods. One example of this philosophy has already been mentioned in the text - the destruction of some of the world's forests because prices were 'right'[7]. When Georgescu-Roegan examined present policies on the production of fossil fuels he noted that the most easily accessible resources were being used now. The most rational action would be 'to minimize future regrets rather than maximizing present satisfaction'[8].

Forecasts

Energy demand forecasts are necessary for governments to plan how their future energy supplies are to be met. The two methods used in the major industrial countries until very recently were a combination of judgement, based on an awareness of political and economic trends, and the projection of previous trends into the future, the 'Time Series Analysis' approach. These methods were quite good for short term predictions but not satisfactory for longer than ten years. From the early 1950s until 1973 the major industrial countries experienced a continuous and increasingly rapid demand for primary energy which accompanied their economic growth. For individual countries the energy elasticity or energy coefficient, defined as the ratio of the annual rate of growth in energy consumption to the annual rate of growth in gross national product, was very stable in this period. The conclusion was drawn that there was a firm and direct relationship between energy use and economic activity. A further assumption was that the energy elasticity experienced in previous years was not going to change very much in the future. This was the position in the UK in 1977 when the National Centre for Alternative Technology (NCAT) invited an independent group, which included the author, to prepare an alternative energy strategy for the UK. We did not accept the traditional assumptions. Our assumptions included that the service sector would continue to grow relative to manufacturing and that industries which made products with a high value/energy index e.g. computers, would expand at the expense of more traditional energy consuming industries such as iron and steel, heavy engineering and bulk chemicals[10]. This resulted in the curved projection shown as NCAT in Figure 10.1, which is compared with a range of official Department of Energy projections at several dates[11,12,13].

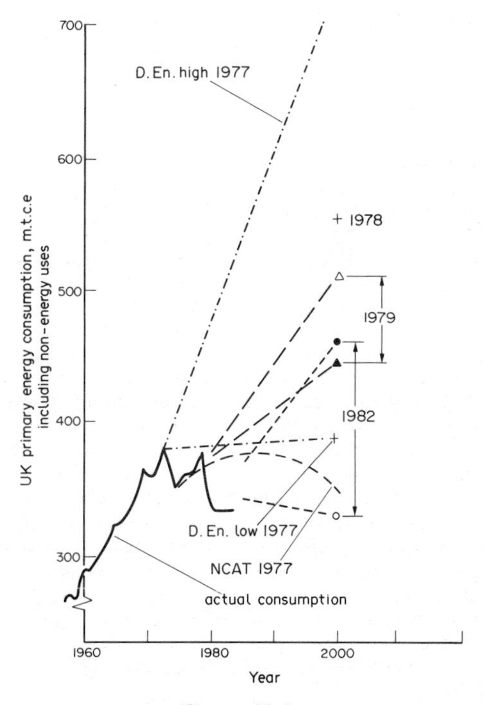

Figure 10.1

Official projections can include options which are highly unlikely to illustrate 'what could happen if..' A good example is the High Growth 1977 estimate in which no real increase in the price of oil was assumed up to the year 2000. Our projection suggested that overall demand for primary energy in the UK would not increase by the year 2000 and would decline thereafter, although we assumed a 2% per annum growth in GNP. Since 1973 there have been very considerable fluctuations in the energy elasticities of most of the industrial countries. There is now general agreement that the former direct relationship between energy use and economic activity no longer exists. The first major report to draw this conclusion was the Saint-Geours Report for the EEC, published in 1979[14]. Calling the slackening of the relationship 'dissociation', the growth in the volume of the gross domestic product and in gross primary energy consumption in the nine Member States of the EEC was examined in the period from 1973 to 1978. For the community as a whole, gross domestic product increased by 10.24%. Primary energy consumption grew by only 0.42%. Dissociation was not the same in all the Member States, as shown in Table 10.2, as they were not all at the same stage of industrial development.

Table 10.2

Comparison of growth in gross domestic product(GDP) and energy consumption 1973-1978

	GDP%	Energy Consumption %
France	+15.03	+0.95
Ireland	+18.89	+2.41
UK	+ 4.35	-3.23
All EEC countries	+10.24	+0.42

Data from reference 14.

Energy forecasts based on the former assumptions of projecting previous trends and assuming the direct relationship between energy use and economic activity, resulted in substantial over estimates of likely future demand. The magnitude of some of these overestimates was startling. In Canada the growth in primary energy consumption was about 5.3% per annum between 1965 and 1973, as shown in Table 2.5. Primary energy consumption in 1973 was equivalent to 7.85×10^{18}J and was projected to reach 19.4×10^{18}J in 1990. By 1980 the National Energy Program had changed the official projection to 11.3×10^{18}J[7]. The seven year period from 1973 to 1980 had resulted in a reduction in projected demand which was actually greater than the total Canadian consumption in 1973.

A third forecasting method using econometric methods has recently been shown to give considerably better results when applied to energy demand forecasts. The first step is to select an econometric equation that relates the explained variable, in this case energy demand, to the explanatory variables, for example GDP and energy price. The equation is then solved to give the best estimate of the parameters. This is straightforward with one or two explanatory variables, but the mathematics is complex for more than two variables and is equivalent to curve fitting on a multi-dimensional space[15]. The econometric equation which fits the energy demand in the UK over the period since 1973 has been found to be[15]

$$\text{Energy demand} = k[\text{GDP}]^a \times (P)^b$$

where P represents the average price of energy, and a, b and k are selected to provide the best fit to the data. The values of a and b were 0.83 and -0.3 respectively.

Energy conservation and some new concepts.

Throughout the world there are probably thousands of research and development projects examining new ideas for obtaining and using energy resources. Some ideas for reducing the amount of energy used in particular applications have been available for many years but have only recently become better known through various energy conservation campaigns. Other projects have reached the pilot plant or pre-production model stage. Several attractive ideas have been given low priority because they would not be cost competitive with other systems. In the following paragraphs a few examples have been selected to illustrate the very wide range of this work.

Deep high-pressure natural gas is thought to be present in such enormous quantities according to Gold[16] that it will be found in most areas of the world and will become the chief fuel for all stationary power use. Gold's theory, which has caused very considerable controversy among geologists, is that the carbon which is known to exist at the earth's surface and in the atmosphere originated as methane. If this is true, the methane could have supplied today's world energy use for some 10 million years. The evidence that there could be vast amounts of methane waiting to be extracted comes from the many reports of sheets of flames associated with earthquakes. The difficulty of proving or disproving the theory is that holes would need to be drilled to very great depths and drilling costs rise exponentially with depth. By 1984 one such hole was being drilled in Sweden. Gold has stated that the global energy shortage is just the result of believing that all likely places have been examined for fuels. If the conventional viewpoint is not correct huge new possibilities would be opened up.

Tidal stream power or the power from river currents has been investigated by the Intermediate Technology Group at Reading University[17]. They estimated that the total power in tidal schemes around the UK averaged 15 GW, and that the power in river currents could be of very considerable significance in some of the developing countries. Their method of harnessing this power is based on the underwater equivalent of the vertical axis windmill. It has a number of advantages in river current energy applications including its high efficiency, low solidity as only a small part of the area swept by the rotor is occupied by blades, and the vertical shaft avoids the need for a right-angled transmission drive. Several small prototypes have been undergoing field trials in Africa.

Salinity gradient power is a very simple concept. If pure water and a saline solution are placed on either side of a semi-permeable membrane, the water molecules will permeate the membrane and dilute the saline solution. This is the process of osmosis and in equilibruim the level of the salt solution will be higher than that of the water. The resulting head of water could then be used to flow through conventional turbines to generate power[18]. The osmostic pressure head differential between ocean and fresh water is equivalent to a column of water 240 metres high, so the concept envisages large dams at the mouths of rivers as they enter the sea. The theoretical potential power is up to 2000 GW if all world rivers are used in the calculation. In practice the development of a suitable membrane is considered to be very unlikely[19].

The modern fuel cell is a new development of an old concept. It is now regarded as one of the most promising new energy conserving developments for the 1990s. It was invented in the early 1840s by Sir William Grove, who wrote to Michael Faraday describing a 'gas battery' in which hydrogen and oxygen reacted against platinum electrodes dipping into dilute sulphuric acid[20]. For about a hundred years it remained a laboratory curiosity until another Englishman, Dr. Francis Bacon, produced the first practical fuel cell and by 1954 a cell producing 150 W had been built at Cambridge

University. Fuel cells directly convert the chemical energy of a fuel into electrical power with an efficiency of 40 to 45%. This overcomes the thermodynamic limitations of conventional power generation equipment with much lower efficiencies.

Another application is in the combined generation of heat and power, where for every 100 units of primary energy consumed by the fuel cell, 40 units of power and 40 units of heat can be produced[20]. In its simplest form hydrogen gas is supplied to a porous electrode which acts as the anode and oxygen is supplied to another electrode which acts as the cathode. The reaction takes place in a potassium hydroxide electrolyte. Because there is no combustion, fuel cells are pollution-free, generating electricity with only air, water and some waste heat as the exhaust products. Fuel cell research continued in the UK until the early 1970s, when commercial development was considered to be unlikely for economic reasons. However in the United States the space programme, particularly for the Apollo missions provided the impetus for new research. The Apollo fuel cells had electrical outputs of 2.5 kW. Between 1974 and 1977 a 1.2 MW demonstration fuel cell for terrestrial applications had been developed and by 1983 two 4.8 MW demonstration power stations in New York and Tokyo were due for evaluation[20]. Terrestrial fuel cells use more readily available fuels, such as natural gas and naptha. Projected costs in large scale production could be in the region of £500/kWe[20].

The heat pump is the reverse of the heat engine, as described in Chapter 1. The performance ratio is Q_H/W, where Q_H is the amount of heat delivered at the higher temperature and W is the work input. Performance ratios of 3 or more can be obtained from commercially available equipment, but the ratio falls with lowered inlet heat source temperature. Most heat pumps are driven by electric motors, so their main advantage lies in overcoming the thermal inefficiency of conventional electrical power generation - electricity delivered with 30% energy conversion efficiency can be upgraded to give an output approaching the original energy potentially available in the fuel. Considerably more energy than that apparently available in the fuel can be obtained if a gas fired engine drives the heat pump[21]. The principles are illustrated in Figure 10.2.

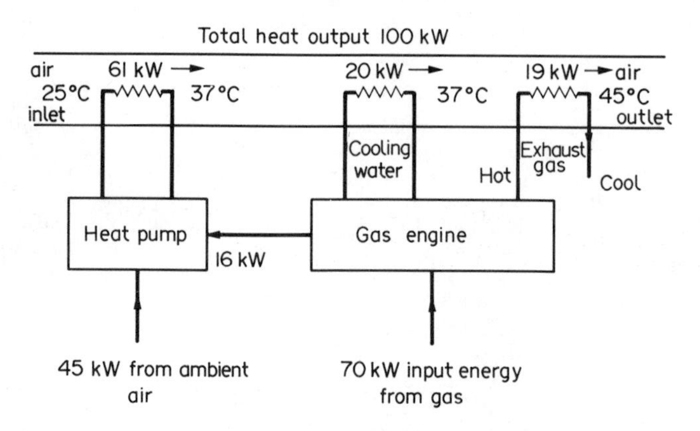

Figure 10.2 Air heating system

During operation the fluid to be heated receives heat from the condenser, the cooling water from the gas engine and finally from the exhaust gases. The useful heat outputs from air to water heating systems could be 1.2 times the fuel energy input, while for air to air systems the ratio reaches 1.42[21]. A number of prototype units of both types were being evaluated in the UK during 1984.

District heating is another simple concept which can greatly improve the overall efficiency of energy conversion into heat. The water is heated in a central boiler and then piped to the individual consumer. This is a well-known technology in many countries, such as Denmark, West Germany, Sweden, Romania and the USSR, where up to about half the space and hot water demand is supplied[22]. In the UK there are only two schemes, one in Nottingham and the other in Peterborough, which are large enough to be classified as district heating schemes, although a number of smaller group schemes have been developed by local authorities or housing associations or the National Coal Board[22]. In addition to the improved overall energy conversion efficiency and reduced running costs from the larger boiler, the possibility of using 'waste' heat from other sources is there, should the opportunity occur.

Combined Heat and Power is the natural development from the district heating concept. Typical figures for electricity generation in steam turbine plant show that only some 30 to 35% of the energy originally available in the fuel is converted into electricity. The remainder appears as hot exhaust gases or as hot water resulting from cooling the turbine condensers. If this waste heat could be supplied to meet the space and water heating requirements of consumers through a district heating network, the overall energy conversion efficiency would be considerably increased. Broadly similar arguments apply to gas turbine or diesel generation, but there are other important considerations. In countries where the winters are not particularly long and cold, such as the UK, the demand for heat would be far less than the optimum heat availability for maximum efficiency. A study for a combined heat and power station in Hereford showed that a gas turbine based scheme could have 21% of the energy converted to electricity and 46% to high grade heat, giving an overall efficiency of 67%. But the scheme would not produce sufficient financial revenue to support the cost of the equipment[23]. The Hereford scheme was able to proceed because there was a large industrial process heat demand. Energy conversion was then biased towards electricity generation by diesel engine, because of the higher themal efficiency, in the order of 40%, and the ability to operate on relatively cheaper fuel. The electrical output was provided by two sets rated at 7.5 MW, with an associated heat output of 15 MW. The designed overall efficiency was 76%. Additional oil-fired boilers can provide extra heat to meet peak demand when necessary. In its first six months of operation it achieved an overall thermal efficiency of 68%. In 1979 a major report on the feasibility of combined heat and power in the UK[24] concluded that annual energy savings equivalent to 30 m.t.c.e. could be possible within some twenty-five years and that the concept deserved a prime place in the country's energy policy.

The car for the nineties is the third energy conservation vehicle(ECV) developed by British Leyland. The first two were small car prototypes. ECV3 is a medium sized family car. Their work started in the late 1970s with an examination of the total energy used in the materials, production and running of cars[25]. Table 10.3 shows that some 85% of the total energy is used on the road.

Table 10.3.
Lifetime Energy Consumption for 100 000 miles

	Materials	**BL factory**	**Fuel at 35 mile/gal**
Gigajoules(GJ)	44	32	450
Cost(£)	88	96	4860

Data from reference 25.

The objectives were to achieve an acceleration from 0-60 mile/h in 12 seconds or less, average fuel consumption to be better than 50 mile/gal, with the normal features of ride, noise levels, handling, safety, life and performance as good or better than the current norm. It must also meet pollution regulations. The design philosophy concentrated on improving the fuel consumption by reducing the weight, improving the engine and power transmission system and reducing the aerodynamic drag. All the objectives were achieved, with a maximum speed in the region of 115 mile/h and the acceleration time to 60 mile/h in just under 11 seconds. The 'typical' fuel consumption was 62.6 mile/gal with 81 mile/gal recorded at a steady 56 mile/h. BL emphasize that this is not a production prototype. The most cost-effective ideas, components, materials and precesses will be applied progressively to production models.

Energy recovery from wastes and the philosophy of effective waste management are rapidly achieving major importance[26]. There are two main reasons for this. The first is to conserve the conventional forms of energy and raw materials, the second is to protect the environment from excessive pollution. Re-usable waste materials should always be recycled, the maximum amount of energy should be recovered from the combustible portion and the residue should then be dealt with in a manner which will cause least damage to the environment. Glass, ferrous and non-ferrous metals are all materials which need a high energy input to produce. In many cases less than half this energy is needed to produce the recycled product. Refuse-derived fuel (RDF) is a fairly dense solid material which must be used locally as transport costs over long distances would exceed the value of the fuel, but use in power stations in the UK is beginning to increase. Typical heating values of RDF as fired are from 7000 to 11,500 MJ per tonne[27]. One system installed in a furniture factory in Sussex saved 65% of the heating costs compared with oil and a further 10% saving on not having to dispose of wood waste as refuse[27]. In the UK the estimated annual total of 30 million tonnes of rubbish has a heating value equivalent to 12 million tonnes of coal, enough to make a small but significant contribution to UK energy supplies.

Energy management is a relatively new discipline which could be defined as the planning, controlling and monitoring of the utilization of energy, to produce the right amount, at the right time, in the right place, with the correct quality as economically as possible. At its simplest level energy management involves the utilization of the minimum amount of energy to perform any particular operation. It is important to match the end-use with the quality of the fuel, a point made in the discussion on first and second law efficiencies in the first chapter. The use of the high quality fuels such as natural gas or oil to generate low-grade heat is a misuse of the potential in the fuel to produce work or generate electricity. Energy management is also concerned with the organizational, financial and managerial functions concerned with overall energy efficiency. Three basic approaches can be followed to reduce energy consumption[28].

(i) Improve efficiency by the elimination of unnecessary use through better housekeeping and operational procedures or more efficient equipment.

(ii) Reduce use by better regulation or by deciding to use less, possibly as a result of economic pressure

(iii) Substitute a better matched energy form.

One example of energy management in a large organization in the UK with over 2000 buildings to control was reported in 1984[29]. The Avon County Council spent £2½ million in the six year period from 1977 to 1983 on a variety of energy conserving measures which included

(i) Fitting automatic controls

(ii) Wall and roof insulation

(iii) Draught-sealing ill fitting windows, doors and roof lights

(iv) Fitting storm porches

(v) Changing to more economical tariffs

(vi) Replacing inefficient lighting sources with new low energy, high output lamps

(vii) Installing energy management zone controls to enable unoccupied parts of the building to be switched off.

A computerized total energy monitoring system was used to check fuel accounts, so that any unusual rises could be investigated. Straightforward good housekeeping was estimated to account for 10% of the total savings. These came to nearly £$8\frac{3}{4}$ million over the six years, giving a net saving of about £$6\frac{1}{4}$ million. Estimates of how much energy could be saved throughout the UK range from 20% to 45% of current (1984) usage. At the level of the individual household in the UK there is evidence that the Government provides an enormous subsidy, in the order of £1400 million, to help pay the fuel bills of those receiving supplementary benefit[30]. This is in contrast to the £18 million allocated to improve the energy efficiency of these households.

Magnetohydrodynamic (MHD) power generation converts very hot combustion gases directly into electricity without the intermediate use of steam or gas turbines. The principle is simple. When a conductor moves perpendicularly to a magnetic field electricity is generated. MHD generators operate at high temperatures, in the order of 2500K. The hot gases contain a seed material, such as potassium, so that it is ionized. The passage of this electricity conducting gas through the transverse magnetic field generates a current at electrodes. The conversion efficiency of a MHD power generator is comparatively low, typically some 25%, but its great advantage is that it can be used in series with a conventional steam power plant as a 'topping up' cycle, with overall efficiences up to 60% with second generation equipment[31]. International cooperation in MHD research has taken place for over 25 years with the leading countries being the Soviet Union, the United States, Australia, India, Japan, the Netherlands, the People's Republic of China and Poland[32].

Summary

'Ultimately the human community must rely on inexhaustible sources of energy'

This brief extract from the Brandt Report[4] should be the major long term objective for a sustainable world energy policy. Perhaps the starting point could be the recognition that a gradual transition to the use of the renewable sources of energy is a desirable long-term aim. This could result in a reassessment of technically promising developments, such as the wave energy programme in the UK, by giving a special economic 'weighting' in any calculation of projected costs of energy produced by the renewable sources. The Science Council of Canada[6] has explicitly argued the need for a long-term planning perspective:

> 'The future has little economic or political power: it has no votes. The government in power, which is a surrogate for the country itself, must take the longer view. It is the responsibility of the government to ensure that future citizens are provided with options-if necessary a trade-off may have to be made against the demands and perceived needs of present citizens'.

Contrast this statement with the UK Government's views on energy policy, which were restated in the Department of Energy's Proof of Evidence for the Sizewell 'B' Public Inquiry[13].

> 'most of the decisions which shape the future level and structure of energy demand are determined by market forces.'

Economic theories which imply a need for increasing the energy consumption of the developed countries must be rejected. The problems of providing sufficient conventional energy sources for the developing countries at a price they can afford must be solved.

The direct link which was assumed to exist between growth in energy consumption and growth in economic activity as measured by GDP has been shown to be no longer true. The developed world and the developing world could both aim for a new target as indicated in Figure 10.3.

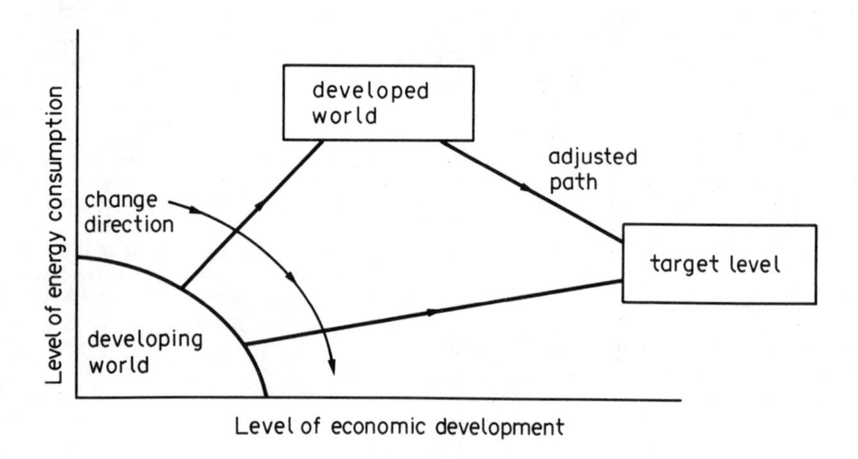

Figure 10.3

A few general observations can be made:

(i) While present proven reserves of oil, natural gas and coal are sufficient for many years at today's consumption levels, continued growth in consumption could result in very rapid depletion of these resources.

(ii) Continuing exponential growth in energy production and consumption or in the economy of any country is therefore not possible.

(iii) Many of the developing countries will need more energy to achieve a stable developed position. Some need immediate help with their acute deforestation problems.

(iv) Problems of increased waste and pollution associated with increased material production and economic activity could be eased by waste management.

(v) The environmental impact of growth in the use of coal and nuclear power could become increasingly unacceptable.

(vi) The period of time necessary to make substantial changes in the pattern of energy supply and demand or for developing new technologies can be long, in the order of at least 15 to 20 years - perhaps up to 50.

There are two precedents, oil and natural gas, for the possibility that at least one of the renewable energy resources or geothermal 'hot rocks' could be providing a significant contribution to the world's energy requirements within fifty years. Figure 10.4 shows the present world annual energy consumption (equivalent to some 45 minutes of solar radiation landing on the earth's surface), with the approximate number of years of ultimately recoverable reserves at today's consumption rates for the fossil fuels. Hydropower could be up to 5 times larger and biomass estimates suggest that 6 times more than today's levels are possible, although the total can only be an informed guess.

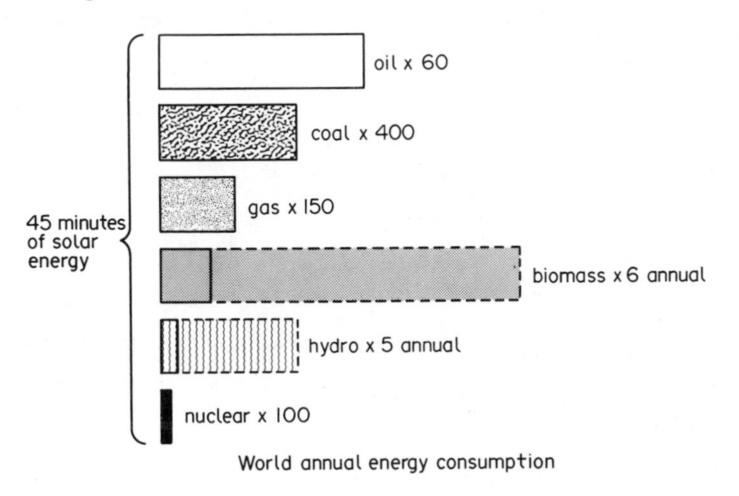

Figure 10.4

Over the next two or three years world energy production and consumption can be estimated with reasonable confidence. Proven reserves of the major fossil fuels do not give any immediate short term cause for concern and production could rise annually by up to three percent per annum to meet any increase in demand following the long-awaited emergence from the period of recession in the early 1980s. Estimates of possible production and consumption patterns during the early 1990s are less certain. The world population figures shown in the projections in Figure 2.4. indicate a level of 5.5. billion. Growth in both hydropower and nuclear energy can be predicted with the schemes already started or planned in the early 1980s. Coal production could continue to increase at the levels experienced in the late 1970s of some $2\frac{1}{2}$% per annum, while the large increased in proven natural gas resources in the USSR could allow an annual increase in gas production of about 3%. Oil production and consumption would probably not be limited by the resources, as proven reserves were sufficient to last until nearly 2020 at the consumption rates experienced in 1982 and 1983, so a modest $1\frac{1}{2}$% per annum growth could be met. This could lead to the hypothetical world energy consumption in 1992 shown in Table 10.4 compared with actual consumption in 1982.

Table 10.4

World Energy Consumption
Million tonnes of oil equivalent

	1982	1992 projection
Oil	2818	3200
Coal	2040	2600
Gas	1370	1800
Biomass+	700	700
Hydropower	446	650
Nuclear	216	650
	7590	9600

+ Estimated

1982 data from reference 33.

The total for 1992 is consistent with the world energy consumption per capita remaining reasonably constant, to allow for a much greater proportionate increase in energy use in the developing countries, but individual totals have been derived from growth rates which may not occur. Any attempt to visualize the situation in fifty years time is an interesting exercise and will be subject to a greater margin of error. Two trends are quite possible. The demand for liquid and gaseous fuels, particularly for transport, will probably be met by the increasing use of synthetic oil and gas, including hydrogen. These could be derived from coal, or possibly from biomass, oil shales and tar sands. The increasing demand for the major fossil fuels will result in considerable technical improvements which will increase the resource base.

The solutions to some of the problems posed in these chapters will require a 'degree of national and international co-operation that the world has not yet seen'[3]. In the long term the problems of waste, pollution and resource depletion associated with the non-renewable resources can only be overcome by a gradual transition to a balanced ecological system.

It will be a difficult path, but one which can be followed when there is a greater understanding of the reasons why the laws of thermodynamics cannot be ignored. Education at every level is the key factor which can influence this move towards less energy-intensive life styles.

> 'Give a man a fish and you feed his family for a day,
> teach him to fish and you feed his family for a lifetime'

References

1. Hubbert, M. King. Energy Resources. A report to the Committee on Natural Resources of the National Academy of Sciences - National Research Council, National Research Council, Washington, D.C., 1962

2. Tiratsoo, E.N., Oilfields of the world, Scientific Press Ltd., Beaconsfield, Bucks, 1973

3. Baxendell, P.B., Energy in the 2000s - what happens to the 'have nots'? Jubilee Lecture, Imperial College of Science and Technology, London, 10th March 1981

4. North-South: A programme for survival, Report of the Independent Commission on International Development Issues, Chairman, Willy Brandt, Pan Books, London, 1980.

5. Eckholm, Erik, Planting for the Future: Forestry for Human Needs, Worldwatch Paper 26, Worldwatch Institute, Washington D.C., 1979

6. Canada as a Conserver Society: Resource Uncertainties and the Need for New Technologies, Science Council of Canada, Report 27, 45, Supply and Services Canada, Ottawa, 1977

7. Schrecker, Ted, The Conserver Society Revisited, Science Council of Canada, Discussion Paper D83/3, Supply and Services Canada, 1983

8. Georgescu-Roegan, N., The Entropy Law and the Economic Process, Harvard University Press, 1971

9. Soddy, Frederick, Matter and Energy, Williams and Norgate, London, 1912

10. Todd, R.W. and Alty, C.J.N. (Eds), An Alternative Energy Strategy for the United Kingdom, National Centre for Alternative Technology, Machynlleth, Powys, Wales, 1977

11. Energy Projections, Department of Energy, London, 1979

12. Energy Policy, A Consultative Document, Cmmd. 7101, HMSO, London, 1978

13. Proof of Evidence for the Sizewell 'B' Public Inquiry, Department of Energy, London, 1982

14. In favour of an energy efficient society, the Saint-Geours Report, a study prepared for the EEC: DG XVII -235 (79) EN., 1979

15. Prosser, R., Forecasting energy demand - the econometric approach, Brighton Polytechnic Lecture, Corporate Strategy Branch, CEGB, London, 1983

16. Gold, T., The Earthquake Evidence for Earth Gas, in Energy for Survival, Ed. H. Messel, Pergamon Press, 1979

17. Fraenkel, P.L. and Musgrove, P.J., Tidal and River Current Energy Systems, Int. Conf. Future Energy Concepts, 114-119, Institution of Electrical Engineers, London, 1979

18. Brin, Andre, Energy and the Oceans, Westbury House, IPC Business Press, 1981

19. Riva, J.P. et al., Energy from the Ocean, Science Policy Research Division, Library of Congress, Washington, D.C., 1978

20. Ryan, F.J. and Cameron, D.S., Fuel Cells- a potential means of energy saving by on-site cogeneration of heat and power, in Energy and our Future Environment, Institute of Energy, London, 1983

21. Hoggarth, M.L., and Pickup, G.A., The role of gas fuelled heat pumps for space heating, in High Efficiency Appliances for Domestic Heating, British Gas Corporation, London, 1983

22. Flood, Michael, Solar Prospects, Wildwood House, London in association with Friends of the Earth, 1983

23. Price, M.E., Hereford Combined Heated Power Station, Int. Conf. Future Energy Concepts, Institution of Electrical Engineers, London, 1981

24. Combined Heat and Electrical Power Generation in the United Kingdom, (W. Marshall, ed.) Energy Paper No. 35, Department of Energy, HMSO, London, 1979

25. King, C.S., A car for the nineties: BL's energy conservation vehicle, Proc. I. Mech. E., 197, 64, 1983

26. Kut, David and Hare, Gerard, Waste Recycling for Energy Conservation, The Architectural Press, London, 1981

27. Hare, Gerard, Private Communication

28. Smith, Craig B., Energy Management Principles, Pergamon Press, 1981

29. Energy Management, Department of Energy, London, January 1984

30. Boardman, Brenda, The cost of warmth, National Right to Fuel Campaign, London, 1984

31. Coal Utilization, UNESCO, Paris, 1983

32. Sluyter, M.M. (Ed.), MHD Electrical Power Generation, UNESCO, Paris 1982

33. B.P. Statistical Review of World Energy 1982, The British Petroleum Company plc, London, May 1983

Further Reading

A broader, non-mathematical approach to energy studies which also explores the social and economic issues is given by Gerald Foley in The Energy Question, Second Edition, Penguin Books Ltd., 1981. Policy, planning and modelling are covered in very considerable depth by The Global 2000 Report to the President of the U.S., prepared by the Council on Environmental Quality and the Department of State, Pergamon Press, 1980. A wide view of the main issues, biased towards the needs of scientific sixth formers and first year university students, is provided by Energy for Survival, edited by H. Messel, Pergamon Press, 1979. Professor Richard C Dorf's Energy, Resources and Policy, Addison-Wesley, 1978, is a major work covering all the main issues, while his shorter paperback The Energy Answer 1982-2000, Brick House, Mass., covers similar ground with a non-mathematical approach. In both books the emphasis is on the United States viewpoint. An important contribution which examines present energy use patterns and shows how economic growth could be sustained without increased energy consumption is given in A Low Energy Strategy for the United Kingdom by Gerald Leach and colleagues at the International Institute for Environment and Development, published by Science Reviews Ltd., London, 1979. Energy Management Principles, by Craig B. Smith, Pergamon Press, 1981, gives a serious introduction to the general methods of energy management while Waste Recycling for Energy Conservation, by David Kut and Gerard Hare, Architectural Press, London 1981, concentrates on waste recycling and management. Among the journals, Pergamon Press publish Solar Energy, Energy and Hydrogen Energy, while the International Journal of Ambient Energy covers all the renewables.

APPENDIX

SOME UNITS AND CONVERSION FACTORS

(a more detailed summary is given on pages 7-10)

Length

1 millimetre (mm)	0.0393701 inch (in)
1 metre (m)	3.28084 feet (ft)

Area

1 square centimetre (cm^2)	0.155000 in^2
1 square metre (m^2)	10.7639 ft^2
1 hectare = $10^4 m^2$	2.4710 acres

Volume

1 cubic centimetre (cm^3)	0.0610237 in^2
1 cubic metre (m^3)	35.31477 ft^3
1 litre (l) (or one thousand cubic centimetres)	1.75985 UK pints
1 imperial gallon (UK)	4.54596 litres
1 US gallon	3.78531 litres

1 barrel = 42 US gallons = 34.97 UK gallons = 159.00 litres
1 barrel of oil = 0.136 tonnes and has an equivalent heat content of 5.694×10^9 J

Weight

1 kilogram (kg)	2.20462 lb
1 tonne (10^3kg)	0.9984207 ton (UK)

1 ton (UK) or Statute or long ton = 1.120 short tons

Force

1 newton (N)	0.2248 lbf

Pressure

1 pascal (Pa)	1 Nm^{-2}
1 bar = 10^5 Pa	14.50 lbf in^{-2}
1 lbf in^{-2} (one pound per square inch or psi)	6.89476 kPa
Atmospheric pressure = 14.70 lbf in^{-2}	101.325 kPa

Heat, Energy and Power

1 watt = 1 Joule per second = 0.001341 horsepower (hp)
1 horsepower = 745.7 Watts = 550 ft lb per second
1 kilowatt (kW) = 1.34 horsepower
1 Btu = 1.05506 x 10^3J = 778.169 ft lb
1 kilowatt hour (kWh) = 3.6 x 10^6J or 3.6 MJ = 3.41213 x 10^3 Btu
1 therm = 10^5 Btu = 29.3 kWh = 1.05506 x 10^8J
1 k cal/m^2 = 0.3687 Btu/ft^2 = 1.163 Wh/m^2
1 W/m^2 = 3.6 kJ/m^2/h = 0.317 Btu/ft^2/h
1 Btu/ft^3 = 3.726 x 10^4 J/m^3

Abbreviations in SI Units

Kilo	k	10^3	milli	m	10^{-3}
Mega	M	10^6	micro	μ	10^{-6}
Giga	G	10^9	nano	n	10^{-9}
Tera	T	10^{12}	pico	p	10^{-12}

For example, 1 megawatt = 1MW or one thousand kilowatts or one million watts.

Temperature Conversions

1 scale degree Celsius (Centigrade), °C = 1.8 scale degrees Fahrenheit, °F.

$$°C = 5/9\,(°F - 32) \text{ or } °F = 9/5\,°C + 32$$

Temperature on the Kelvin Scale is obtained by adding 273.15 to temperatures on the Celsius Scale. Note that 40°C-20°C = 20K.

Conversion Factors for the Fossil Fuels

Coal — 1 million tonnes is equivalent to 0.6 million tonnes of oil or 250 million therms (2.5 x 10^{13}Btu) or 7.35 TWh electrical energy or 2 TWh elecricity generated, or 2.637 x 10^{16}J.
(m.t.c.e. is "million tonnes of coal equivalent")

Oil — 1 million tonnes is equivalent to 1.7 million tonnes of coal or 425 million therms (4.25 x 10^{13}Btu) or 12.5 TWh electrical energy or 3.6 TWh electricity generated, or 4.484 x 10^{16}J. The definition given in the BP Statistical Review of World Energy 1982 equates 1 million tonnes of oil with 3.97 x 10^{13}Btu and 1.5 million tonnes of coal equivalent, or 4.0 TWh electricity generated, or 4.115 x 10^{16}J.
(m.t.o.e. is "million tonnes of oil equivalent")

Natural Gas — 1 million cubic feet is equivalent to 40 tonnes of coal or 10 thousand therms. A common unit is a trillion cubic feet (Tcf) which is 10^{12} cubic feet, and equivalent to 40 million tonnes of coal (1.05 x 10^{18}J). 1 cubic foot is the equivalent of 1000 Btu or 1.05 x 10^6J. This gives 1 cubic metre as the equivalent of 3.71 x 10^7J. The BP Statistical Review prefers 1 cubic metre as the equivalent of 9000 k cal, which is 3.77 x 10^7J.

GLOSSARY

Absorptivity The ratio of the amount of radiation absorbed by a body to the radiation incident upon it. (absorptivity = 1 - reflectivity.)

Acid precipitation Often called 'acid rain', it is largely caused by the interaction of sulphur dioxide and nitrogen oxide emissions with water in the atmosphere.

Aerobic fermentation Fermentation caused by micro-organisms (bacteria) in the presence of oxygen.

Algae Filamentous or unicellular water plants, normally fast growing.

Anaerobic fermentation Fermentation caused by micro-organisms (bacteria) in the absence of oxygen.

Anthracite Coal with a calorific value greater than 32.5 MJ kg^{-1}, containing 95-98% carbon.

Associated gas Natural gas found in sedimentary rock reservoirs associated with petroleum.

Atom The smallest particle of any element which has the chemical properties of that element.

Atomic Number Determines the chemical properties of the element.

Attenuator A long thin wave energy converter structure which is aligned parallel to the incident wave direction.

Available Work A concept which describes the quality of any energy transformation as well as its quantity.

Available Energy The maximum useful work that can be performed by a system interacting with the environment, also known as 'exergy'.

Barrel A standard volumetric unit equivalent to 42 US gallons or 159 litres.

Bioconversion The conversion of solar energy into chemically stored energy through biological processes. Various fuels and materials can be produced by bioconversion.

Biogas A flammable gas produced by anaerobic fermentation. In chemical composition it is a mixture of some 60-70% methane, with carbon dioxide and traces of other gases.

Biomass All types of animal and plant material which can be converted into energy.

Bituminous coal Coal with a calorific value greater than 26.7 MJ kg^{-1}, containing 82.5-92% carbon and between 4.2 and 5.6% hydrogen.

Black body A term denoting an ideal body which would absorb all and reflect none of the radiation falling upon it. Its reflectivity would be zero and its absorptivity would be 100%. An alternative definition is a body which at any one temperature emits the maximum possible amount of radiation, i.e. its emissivity is 1.0. The total emission of radiant energy from a black body takes place at a rate expressed by the Stefan-Boltzmann Law.

Boiling Water Reactor A nuclear reactor which uses ordinary water as the moderator.

British Thermal Unit The amount of heat required to raise one pound of water through one degree Fahrenheit.

Brown coal Coal with a calorific value less than 23.76 MJ kg^{-1}. In some countries, including Canada and the United States, brown coal has an upper limit of 19.3 MJ kg^{-1}.

Calorie One calorie is the amount of heat required to raise one gram of water through one degree Celsius (Centigrade). A more common unit is the kilocalorie (kcal), often written as 'Calorie'. One calorie is equivalent to 4.1868 Joules.

CANDU A pressurized heavy water reactor developed in Canada. CANDU is derived from Canada, Deuterium and Uranium.

Carnot cycle A theoretical heat engine cycle derived by the Frenchman Sadi Carnot in 1824. The theoretical Carnot cycle efficiency will always be greater than the efficiency of any real engine working between the same temperature limits.

Chain reaction A continuous splitting process caused when neutrons emitted during the fission process strike the nucleus of another uranium-235 atom, releasing more energy and emitting more neutrons.

Chemical Energy Energy stored in a substance due to its chemical composition, usually released by combustion.

Collector A solar collector or absorber is used to collect solar radiation. In the process, the radiation undergoes a change in its energy spectrum.

Concentration ratio The ratio of the irradiance at the focus of the concentrator to the direct radiation received at normal incidence on the surface.

Concentrator A device for focusing solar radiation.

Condensate Liquid natural gas, also known as NGL (natural gas liquid).

Conduction The transfer of heat from one body, or part of a body, to another part of the body or to another body in physical contact with it.

Convection The transfer of heat within a fluid through the internal movement of the fluid. Convection can be natural and caused by differences in density, or forced and caused by a machine such as a pump or fan.

Core The heart of a nuclear reactor containing the fuel.

Crude Oil Petroleum liquid obtained directly from the oil reservoir.

Declination Solar declination is the angular distance between the sun and the plane of the celestial equator.

Delivered Energy Energy actually supplied to the consumer. It is always less than the original primary energy because of transmission and conversion losses.

Deuterium An isotope of hydrogen whose nucleus contains one neutron and one proton, sometimes known as 'heavy hydrogen'.

Diffuse radiation or insolation Solar radiation which arrives on Earth as a result of the scattering of direct solar radiation by water vapour and other particles in various layers of the atmosphere. It is also known as indirect radiation or sky radiation.

Direct Hydrogenation This occurs when a reducing gas, such as high-purity hydrogen, reacts with coal at high temperatures and pressures in the presence of a catalyst.

Direct radiation or insolation Radiation from the sun which eventually reaches the Earth's surface without being scattered. This is also known as direct beam radiation.

Dry Gas A natural gas which contains less than one litre of extractable liquid hydrocarbons per 63 cubic metres of gas.

Emissivity The ratio of the radiation emitted by a body to that which would be emitted by an ideal black body the same temperature and under identical conditions.

End Use Energy See Useful Energy

Energy The capicity to do work. Energy can be transformed from one form to another e.g. heat and work, or interconverted with mass. It can be exchanged from one body to another.

Energy Conversion The transformation of energy from one form to another form.

Enrichment The process of increasing the percentage of an isotope present in a natural nuclear fuel e.g. with uranium the level of fissionable uranium-235 is increased from 0.7 per cent up to three per cent.

Entropy A quantity defined by the German physicist Clausius. In any process the change in entropy is equal to the change in heat divided by the absolute temperature.

Exergy See Available Energy.

Exponential growth A function y of time t, such that the rate of growth is proportional to y (see pages 11-13).

Evacuated tubular collector A collector which has two outer concentric glass tubes. A third central tube contains the heat transfer fluid, while the evacuated space between the outer two tubes reduces heat losses. Characteristically, their efficiency is in the order of 50% with temperature differences between the heated fluid and the surroundings greater than 100°C.

Fast Breeder Reactor A class of reactor which uses fast neutrons with a uranium based fuel which contains up to 30% of plutonium-239, and which is capable of breeding new fissile material.

Fission The splitting of a heavy nucleus into two smaller, approximately equal parts with the release of a relatively large amount of energy and some neutrons.

Flat plate collector Any non-focusing flat surfaced solar collector.

Fluidized-bed A device used in coal-burning power plants in which small solid particles behave as if they were a liquid. Their action is caused by high pressure air movement.

Fossil fuel Any natural organic fuel e.g. coal, natural gas, or any form of petroleum.

Fuel cell An electrochemical device for converting chemical energy directly into electrical energy without combustion.

Fusion The combining of the atomic nuclei of very light elements to form heavier elements with the release of a relatively large amount of energy. The very light elements are usually forms of hydrogen, the heavier element is usually helium.

Geothermal energy The heat energy available at or near the earth's surface in rocks or in the form of hot water and steam.

Geothermal aquifer A source of useful hot water sometimes found at depths between 1.5 and 2 km.

Global radiation The sum of the intensities of the vertical component of direct solar radiation and the diffuse radiation. An alternative definition is radiation at the Earth's surface from both sun and sky.

Greenhouse effect An expression given for the solar heating of bodies shielded by glass or other transparent materials which transmit solar radiation but absorb the greater part of the radiation emitted by the bodies.

Gross Domestic Product (GDP) The current market value of all the various activities within a country, including the production of all types of manufactured goods and services.

Gross National Product (GNP) The sum of the Gross Domestic Product (GDP) and all earnings from foreign investments.

Half-life, radioactive The time required for any radioactive substance to lose 50 per cent of its activity by the transformation of 50 per cent of its atoms to another form. Half-lives can range from a few seconds to thousands of years.

Heat A form of energy. It can be transferred from a body at a higher temperature to another body at a lower temperature as a result of the temperature difference between the two bodies.

Heat engine A system operating in a cycle and producing useful work by exchanging heat with a higher temperature heat source and a lower temperature heat sink. A conventional 'real' heat engine, such as a diesel engine, is not a heat engine in this thermodynamic sense because real engines experience mass flow across their system boundaries.

Heat pipe A device for transferring heat by means of the evaporation and condensation of a fluid in a sealed system.

Heat pump A reversed heat engine; it transfers heat from a lower temperature to a higher temperature by the addition of work. The amount of heat delivered at the higher temperature divided by the work input is called the coefficient of performance.

Heliostat A mobile array of mirrors used to reflect a beam of sunlight in a fixed direction as the sun moves across the sky.

Heavy crude oil Oil with a greater number of larger molecules and a higher specific gravity.

Heavy water Water in which all the hydrogen atoms have been replaced by deuterium atoms.

Hydrocarbon A chemical compound containing only carbon and hydrogen. Hydrocarbons form the main part of the fossil fuels.

Hydroelectric power Electrical power obtained from the convertion of the potential energy of water at a high level into kinetic energy. This is converted into mechanical power in a turbine which converts the mechanical power into electricity.

Hydrogenation, direct See Direct Hydrogenation.

Hydrogen economy A complete energy system in which hydrogen forms the intermediary link between the primary energy sources and the energy consuming sectors.

Incidence angle Angle between the perpendicular to the surface and the direction of the sun at that instant.

Indirect Liquifaction Also known as the synthesis process, it is a method for producing liquid fuels, alcohols and chemicals from coal.

Infra-red radiation The band of electromagnetic wavelengths lying between the extreme of the visible region (circa 0.76 μm) and the shortest microwaves (circa 1000 μm).

Insolation Originally defined as one of the processes of weathering, it is now generally regarded as another term for solar radiation, including the ultra-violet and visible infra-red radiation regions. Total insolation is a term sometimes used instead of global radiation.

Internal Energy The intrinsic energy of a body or a fluid at rest. It can always be measured by referring to other measurable properties such as temperature and pressure.

Irradiance Radiant energy passing through unit area per unit of time.

Irradiation The process of exposing to radiation.

Isotopes The name given to different types of chemically similar atoms with the same number of protons but with different numbers of neutrons.

Joule The fundamental unit of work. One joule is exactly one newton-metre.

Kerogen The solid organic hydrocarbon wax material contained in oil shale.

Kinetic Energy Energy due to the motion of a body of a given mass. The kinetic energy of a body of mass m and velocity u is $\frac{1}{2}\ mu^2$.

Langley A unit of energy frequently used in solar radiation work, equivalent to one calorie per square centimetre.

Lean Gas A natural gas whose extractable liquid hydrocarbons lie between those of a wet gas and a dry gas.

Light crude oil Oil with a smaller number of larger molecules and a lower specific gravity than the heavier crude oils.

Light Water Reactor A nuclear reactor in which ordinary water is the primary coolant. The two commercial light-water reactors are the boiling water reactor (BWR) and the pressurized water reactor (PWR).

Lignite A low-grade coal, with a calorific value lying between the brown coals and the subbituminous coals, often less than 19.3 MJ kg^{-1}.

Liquifaction, coal The conversion of coal into liquid hydrocarbons and other chemical compounds by hydrogenation.

Magnetohydrodynamic (MHD) generator A device for generating electricity from very hot partially ionized gases which flow through a magnetic field.

Magnox Reactor A British gas-cooled nuclear reactor. Magnox is the name given to the fuel cladding which is made from a magnesium alloy.

Methane The lightest of the hydrocarbon gases (CH_4). It is the major constituent of natural gas and is also formed from the decomposition of organic material.

Moderator A material used in a nuclear reactor to slow down neutrons.

Natural Gas Naturally occuring gaseous mixtures. The main constituent is methane, with smaller quantities of ethane, propane, butanes, pentanes and hexanes.

Net Energy Ratio (NER) The ratio of the energy outputs from a system to the energy inputs.

Nuclear Reactor A device for producing heat that can be used for generating electricity through the initiation and maintainance of a controlled fission chain reaction.

Oil Shale A sedimentary rock containing solid organic matter (kerogen). Oil can be obtained by heating the oil shale to high temperatures.

Organization of Petroleum Exporting Countries (OPEC) A group of thirteen countries which hold between them approximately 65% of the world's proven oil reserves. Formed in September 1960, the membership in 1983 consisted of Saudi Arabia, Kuwait, United Arab Emirates, Qatar, Iran, Iraq in the Middle East, Algeria and Libya in North Africa and Ecuador, Gabon, Indonesia, Nigeria and Venezuela.

Ocean Thermal Energy Conversion (OTEC) A device for generating electricity by using the surface water as the upper temperature heat source and the deep ocean waters as the lower temperature heat sink.

Peat Regarded in geological terms as 'young coal'. Hand cut peat has a high moisture content, typically up to 40%, and a calorific value from 11 to 15 MJ kg^{-1}.

Perihelion The point in the earth's orbit when the earth is closest to the sun.

Petroleum A general term describing a wide range of hydrocarbons from natural gas, through crude oils of increasing viscosity to bitumen and solid paraffin waxes.

Photobiology A biological subject covering the relationship between solar radiation, mainly in visible wavelengths, and biological systems.

Photochemistry A chemical subject dealing with chemical reactions induced by solar radiation.

Photosynthesis The conversion of solar energy by various forms of plant and algae into organic material (fixed energy).

Photovoltaic cell Also known as photocell or solar cell. A semiconductor device which can convert radiation directly into electromotive force. An alternative definition is a device used for detection and/or measurement of radiant energy by the generation of an electrical potential.

Plasma A gaseous mixture of electrically neutral positive and negative ions at very high temperatures.

Potential Energy Energy possessed by a body at some height above a fixed datum line. Also known as 'gravitational potential energy', the potential energy of a body of mass m at a height z above the datum line is mgz.

Power The rate of work or the rate at which any energy conversion takes place. The fundamental unit of power is the watt (W). One watt is one joule per second.

Power tower or solar tower A tall tower, perhaps 500 m high, positioned to collect reflected direct solar radiation from an array of heliostats. The top of the tower contains the heat exchange chamber and the hot working fluid is used in a conventional electrical generating system at ground level.

Pressurized Water Reactor A nuclear reactor which uses ordinary water at very high pressure as the moderator.

Primary Energy Energy available in any fossil fuel such as oil or coal before it has been processed or transmitted. It is also the energy potentially available from the natural environment.

Proved Reserves See Reserves (proved).

Pumped storage A system connecting two water reservoirs at different levels. The water can either be pumped from the lower reservoir to the upper reservoir (normally at times when there is surplus electricity available from other sources in the grid system) or can generate hydroelectricity by flowing in the reverse direction. The pump-turbine system is reversible.

Pyranometer An instrument used for measuring global radiation, also known as a solarimeter.

Pyrheliometer An instrument used to measure the direct irradiance of the sun along a surface perpendicular to the solar beam. Diffuse radiation is excluded from the measurement.

Pyrolysis Heating to very high temperatures in the absence of oxygen.

Quad A unit for measuring large amounts of heat energy. One quad is a quadrillion (10^{15}) British thermal units.

'R' factor R is the thermal resistance expressed as the temperature difference required for one watt to pass through one square metre per hour.

Radiant energy Energy transmitted as electromagnetic radiation.

Radiation Radiant energy.

Rankine Cycle The ideal thermodynamic cycle using a two-phase (liquid and vapour) working fluid. Water-steam is commonly used in real conventional vapour power cycles. Organic fluids are used for comparatively small temperature differentials such as those experienced with solar ponds.

Reflectivity The ratio of the radiation reflected from a body to the radiation incident upon it. (reflectivity = 1 - absorptivity.)

Renewable Energy Energy from sources that are continuous or that can replenish themselves naturally e.g. solar energy, wind power, hydropower or biomass.

Reserves (proved) The amount of a known mineral deposit (or fossil fuel) that has been carefully measured and could be extracted economically using existing available technology.

Retrofitting Erecting a solar collecting system on to existing buildings.

Scattering Interaction of radiation with matter where the direction is changed but the total energy and wavelength remain unaltered. An alternative definition is the attenuation of radiation other than by absorption.

Selective surface A surface which has a high absorptivity for incident solar radiation but also has a low emissivity in the infra-red region.

Semiconductor An electronic conductor whose resistivity lies in the range 10^{-2} to 10^{9} ohm-cm (between metals and insulators).

Solar constant The amount of solar radiation which is received immediately external to the Earth's atmosphere and incident upon a surface normal to the radiation taken at the mean Earth-sun distance. It is not a true constant as it varies, mainly due to sun-spot activity. A mean value of 1.373 kW m^{-2} ± 2% is normally taken, although values outside these limits can occasionally occur.

Solar furnace A device for achieving very high temperatures by the concentration of direct radiation.

Solarimeter Another name for Pyranometer.

Solar pond An artificially enclosed body of water containing a stratified salt solution. Solar energy can be stored as heat in the pond, as the stratified salt solution reduces heat losses. Electricity can be generated by using the high temperature solution at the bottom of the pond as the heat source and the lower temperature surface water as the heat sink.

Sour Oil An oil containing more than 0.5% by weight of sulphur compounds, also known as 'high sulphur'.

Spectrometer An instrument for the measuring of radiation intensity over small wavelength intervals.

Sweet Oil An oil containing less than 0.5% by weight of sulphur compounds, also known as 'low sulphur'.

Synfuel A high quality fuel, usually gas or oil, derived directly from one of the fossil fuels. The synfuel is in a more useful form than the original source, e.g. oil derived from coal. However, some of the original primary energy is lost in the conversion process.

Tar Sands Sedimentary rocks containing heavy, viscous petroleum (bitumen) which can be recovered by combustion or other in situ methods as a crude oil.

Thermosyphon Natural liquid circulation caused by the small difference in density between a hot and a cold liquid. In a solar collector thermosyphon system, the collector is placed below the water storage tank and the solar heated water rises to the top of the tank, displacing colder water from the bottom of the tank to the bottom of the collector.

Tokamak A type of nuclear fusion reactor system. The word is derived from a Russian acronym for a toroidal magnetic chamber.

'U' value The reciprocal of the 'R' factor.

Ultra-violet radiation The band of electromagnetic wavelengths lying next to the visible violet (0.10-0.38 μm).

Useful Energy Energy available to the consumer. It can take many forms including heat, light or mechanical work.

Visible region The range between the ultra-violet and infra-red regions. This region can also be defined as that which affects the optic nerves (0.38-0.76 μm).

Wet Gas A natural gas which contains more than one litre of extractable liquid hydrocarbons per 21 cubic metres of gas.

Zenith angle The angle between a line from the sun to the centre of the Earth and the normal to the surface of the Earth (90° - altitude).

INDEX